Progress in Theoretical Computer Science

Howard Karloff

Linear Programming

Birkhäuser   1991
Boston • Basel • Berlin

Howard Karloff
Department of Computer Science
The University of Chicago
Chicago, IL 60637

Library of Congress cataloging-in-publication data

Karloff, Howard, 1961-
    Linear programming / by Howard Karloff.
        p.   cm. -- (Progress in theoretical computer science)
    Includes bibliographical references and index.
    ISBN 0-8176-3561-0
    1. Linear programming.   I. Title.   II. Series.
T57.74.K37 1991   91-3387
519.7 '2--dc20   CIP

Printed on acid-free paper.
© Birkhäuser Boston 1991

ISBN 0-8176-3561-0
ISBN 3-7643-3561-0

Camera-ready text prepared in LaTeX by the author.
Printed and bound by Edwards Brothers, Inc., Ann Arbor, Michigan.
Printed in the U.S.A.

9 8 7 6 5 4 3 2 1

Contents

Preface

Over the more than four decades that have elapsed since 1947, when the Simplex Algorithm was created by George Dantzig, a voluminous theory of linear programming has been developed. This book is an attempt to present a small fraction of this theory to a "mathematically sophisticated" reader. Here, a "mathematically sophisticated" reader is an advanced undergraduate or graduate student who knows linear algebra and who has the ability to read and understand proofs. Except for a few exercises left to the reader, almost all details are included. I have tried to provide intuition and motivation as well.

I have made no attempt to include everything known, or even everything which is important. I have included what I feel every "literate" theoretical computer scientist (or mathematician) should know about linear programming. (Maybe a bit more.) I hope this short monograph will function as a self-contained, concise mathematical introduction to the theory of linear programming.

In my first-year graduate course in the computer science department at the University of Chicago, I usually cover the material here in roughly nine weeks. Any extra time could be spent on the newer polynomial-time linear programming algorithms of, say, Monteiro and Adler [42], Renegar [45], Vaidya [60] or Gonzaga [26]. An alternative is to study the basis reduction algorithm of Lenstra, Lenstra, and Lovász [27, 47]. Yet another possibility is to study the applications to combinatorial optimization of the Ellipsoid Algorithm. The books by Grötschel, Lovász, and Schrijver [27] and by Schrijver [47] contain enough material to keep one busy for far longer than one term.

I am pleased to acknowledge help I had in writing this monograph. Richard Carnes typeset the first draft in LaTeX. Without Richard's superb work, this would be the first handwritten volume of *Progress in Theoretical Computer Science*. Without Sam Webster's help, the figures would look far less professional. Stuart Kurtz and Lance Fortnow indulgently answered my repeated LaTeX questions. The comments of an anonymous reviewer substantially improved the presentation. I am grateful for his or her help. I thank David Shmoys, Ron Shamir and James Renegar for their help in compiling the bibliography. Linus Schrage and Milt Gutterman provided prototypes of large linear programs which are solved in practice. Sundar Vishwanathan helped check some of the proofs, and Steve Fenner simplified the proof of correctness of the Simplex Algorithm. Most of all, I thank my CS 370 (Linear Programming) students from my autumn, '86, '87 and '88 classes, for using an earlier draft of this book as the text.

Howard Karloff

Chapter 1

The Basics

1.1 Introduction

Linear Programming is the process of minimizing a linear *objective function* subject to a finite number of linear equality and inequality constraints. When airlines schedule their crews, when factory managers compute the mix of raw materials which produces the most profitable blend of output products, when utilities plan their oil and gas purchases, they often find themselves setting up and solving problems with hundreds or thousands of variables and constraints. Few problems studied in computer science have greater application in the real world.

Linear programming also has numerous applications within theoretical computer science. It can be used to solve many diverse combinatorial problems seemingly unrelated to linear programming. The Ellipsoid Algorithm for linear programming can be used to give theoretically-fast algorithms for many combinatorial problems, such as finding a maximum flow in a network, finding a maximum matching in a graph, and coloring a perfect graph. In a few cases, the only fast algorithm known is based on the Ellipsoid Algorithm.

Let us start with an example of a linear program, the famous *diet problem*. Imagine that your culinary repertoire is quite limited;

1

you can prepare only four dishes: trout, a corned beef sandwich, a burrito, and a hamburger. In order to minimize decision-making and preparation time, you decide to prepare the same quantity of each every day (good luck!). But you must consume every day at least the minimum daily requirement of vitamins, and in order to maintain your weight you want to minimize your caloric consumption. At the risk of oversimplifying the world, let us assume that your entire nutritional requirements consist of only three vitamins, A, C, and D. This table lists the (entirely fictitious) nutritional value, in milligrams, of each food, the caloric content of the four items you can prepare, and the minimum daily requirements of the three vitamins.

	Vit. A	Vit. C	Vit. D	Calories
Trout	203	92	100	600
CB sandwich	90	84	230	350
Burrito	270	80	512	250
Hamburger	500	90	210	500
Requirements	2000	300	430	

Among all combinations of trout, corned beef sandwiches, burritos, and burgers that satisfy your nutritional requirements, which one minimizes the number of calories consumed? (We allow fractional quantities of each entree, but none can be negative, of course.) To solve this problem, you would naturally create four variables: x_T, the number of trout dishes you would eat in the optimal diet; x_{CB}, the number of corned beef sandwiches consumed in the optimal diet; x_{BUR}, the number of burritos; and x_{HB}, the number of hamburgers. Our goal is to minimize the total number of calories consumed, $600X_T + 350X_{CB} + 250X_{BUR} + 500X_{HB}$, while satisfying the three nutritional requirements. We write three inequalities, one for each of the nutritional requirements. Consider vitamin A, for example. Since each serving of trout has 203 milligrams of vitamin A, each corned beef sandwich, 90 milligrams, each burrito, 270, and each hamburger, 500, we write down

$$203X_T + 90X_{CB} + 270X_{BUR} + 500X_{HB} \geq 2000.$$

Last, we disallow negative solutions by adding the constraints $X_T \geq 0$, $X_{CB} \geq 0$, $X_{BUR} \geq 0$, and $X_{HB} \geq 0$. Here is the resulting *linear program*: Minimize

$$600X_T + 350X_{CB} + 250X_{BUR} + 500X_{HB}$$

subject to:

$$
\begin{array}{rcl}
203X_T + 90X_{CB} + 270X_{BUR} + 500X_{HB} & \geq & 2000 \\
92X_T + 84X_{CB} + 80X_{BUR} + 90X_{HB} & \geq & 300 \\
100X_T + 230X_{CB} + 512X_{BUR} + 210X_{HB} & \geq & 430 \\
X_T \geq 0 \;\; X_{CB} \geq 0 \;\; X_{BUR} \geq 0 \;\; X_{HB} \geq 0 & &
\end{array}
$$

The optimal values of the variables are likely to be nonintegral.

Our second example deals with bipartite graphs. A *bipartite graph* is a pair of finite sets A and B (the *node* sets) and a subset E of the pairs $\{a, b\}$, $a \in A$, $b \in B$. Typically bipartite graphs are displayed pictorially with the elements of E (known as *edges*) represented by line segments, the edge $\{a, b\}$ represented by a line segment between a and b, its *endpoints*. For example, here is a bipartite graph with $A = \{a_1, a_2, a_3, a_4\}$ and $B = \{b_1, b_2, b_3, b_4\}$. The nine edges are arbitrarily labeled $e_1, e_2, \ldots, e_9$.

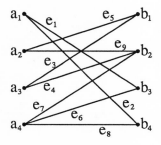

In a bipartite graph, a *matching* is a collection of edges, no two sharing an endpoint. For example, $\{e_1, e_5, e_7\}$ is a matching in the bipartite graph above. A *maximum matching* is a matching with as many edges as possible. Let us attempt to formulate the maximum matching problem as a linear program. We create a variable x_i for edge e_i, whose value is either 0 or 1: if x_i is 1, we include e_i in the matching, if x_i is 0, we don't.

Let us begin by writing the constraint $x_i \geq 0$ for all i; no edge can be chosen a negative number of times. When is a solution a valid matching? If, and only if, there is at most one edge incident to (or "touching") each node. We therefore write a constraint for each node, limiting the number of edges incident to that node to at most one. For example, since node b_2 is incident to edges e_9, e_4 and e_7, we write the constraint

$$x_9 + x_4 + x_7 \leq 1.$$

These constraints imply, in light of the nonnegativity constraints $x_i \geq 0$, that each x_i is between 0 and 1. Though we'd like to stipulate that x_i is either 0 or 1 (and not 0.5, for example), we cannot; let's worry about this later.

Of course, since our goal is to maximize the number of edges chosen, we want to maximize $x_1 + x_2 + x_3 + \cdots$, or minimize $-x_1 - x_2 - x_3 - \cdots$.

Here's the linear program for the example graph above:

	Minimize	$-x_1 - x_2 - \cdots - x_9$			
s.t.	$x_1 + x_2$	$\leq$	1	$x_5 + x_3$ $\leq$ 1
	$x_5 + x_9$	$\leq$	1	$x_9 + x_4 + x_7$ $\leq$ 1
	$x_3 + x_4$	$\leq$	1	$x_1 + x_6$ $\leq$ 1
	$x_7 + x_6 + x_8$	$\leq$	1	$x_2 + x_8$ $\leq$ 1
	$x_1 \geq 0$	$x_2 \geq 0$	$\cdots$	$x_9 \geq 0$

In fact, we need do nothing to ensure that each x_i is either 0 or 1. It is a theorem that in a linear program modeling the maximum matching problem in a bipartite graph, there is always an optimal solution whose entries are either 0 or 1! By virtue of this theorem, we get integrality for free. (But we must find, not just any optimal solution to the linear program, but an optimal one whose components are 0 or 1.)

Formally, a linear program in general form is the problem of minimizing a linear function subject to a finite number of equality and inequality constraints. Where $x, c \in \mathbb{R}^n$, A is a $k \times n$ matrix, $b \in \mathbb{R}^k$, A' is $l \times n$, and $b' \in \mathbb{R}^l$, the following is a general linear program with n variables, k equality constraints, and l inequality constraints. The redundant notation $x_j \gtreqless 0$ means that variable x_j is not sign-constrained. Either k or l can be 0.

$$
\begin{array}{rl}
 & \min c^T x \\
\text{subject to} & Ax = b \\
 & A'x \geq b' \\
x_1 \geq 0 \quad x_2 \geq 0 \cdots x_r \geq 0 & x_{r+1} \gtreqless 0 \cdots x_n \gtreqless 0
\end{array}
$$

Here is an example of a linear program in general form:

$$
\begin{array}{rrl}
\min & 2x_1 - 3x_2 + x_5 & \\
\text{s.t.} \quad -x_1 + 3x_2 + x_3 & = & 12 \\
-5x_3 + 6x_4 - x_5 & = & 19 \\
x_2 - x_4 & \geq & 20 \\
x_1 \geq 0 \ x_2 \geq 0 \ x_3 \gtreqless 0 \ x_4 \gtreqless 0 \ x_5 & \gtreqless & 0
\end{array}
$$

This book focuses on the theory involved in the solution of linear programs.

Let us look at two special classes of linear programs.

Standard form: $\min c^T x$,   in which every variable is sign-con-
s.t. $Ax = b$
$x \geq 0$
strained and only equality constraints are allowed, and

Canonical form: $\min c^T x$,   a form which allows only inequal-
s.t. $Ax \geq b$
$x \geq 0$
ity constraints and in which all variables are sign-constrained. All three forms are equivalent in that any one can easily and quickly be converted to any other. Clearly, a linear program in standard or

canonical form is already in general form, so to prove equivalence it suffices to show that general-form linear programs can be converted to standard and canonical form.

1. General $\rightarrow$ standard: we must eliminate inequality constraints and unsign-constrained variables.

    a. Let $d_i^T x \geq r_i$ be an inequality constraint. Create a new "surplus variable" s_i and replace the inequality $d_i^T x \geq r_i$ by $d_i^T x - s_i = r_i$ and $s_i \geq 0$.

    b. If $x_j \gtrless 0$, create two new variables x_j^+ and x_j^- and replace x_j everywhere by $x_j^+ - x_j^-$, and add the sign constraints $x_j^+ \geq 0$, $x_j^- \geq 0$.

2. General $\rightarrow$ canonical: we must eliminate equality constraints and unsign-constrained variables.

    a. Let $d_i^T x = r_i$ be an equality constraint.   Replace the equality constraint by a pair of inequality constraints

    $$d_i^T x \geq r_i, \quad -d_i^T x \geq -r_i.$$

    b. If $x_j \gtrless 0$, create two new variables x_j^+ and x_j^- and replace x_j everywhere by $x_j^+ - x_j^-$, and add the sign constraints $x_j^+ \geq 0$, $x_j^- \geq 0$.

It is easy to see that the resulting problems are equivalent to the original problem. For the Simplex Algorithm we will use standard form almost exclusively.

1.2   Computational Model

We need some simple notation.

Definition. If f, g are functions from $\mathbb{R}^+$ to $\mathbb{R}$ or from $\mathbb{N}$ to $\mathbb{R}$, we say f is $O(g)$ if there are constants c and x_0 such that for all $x \geq x_0$, $|f(x)| \leq c|g(x)|$.

Let us study our computational model. We will be using a random-access computer with exact addition, subtraction, multiplication, and division of reals. It can also extract square roots exactly and compute $\lceil x \rceil$ of a real x ($\lceil x \rceil$ is the smallest integer greater than or equal to x). Comparisons are allowed and are performed exactly. Each operation takes unit time, independent of the sizes of the operands—1,000,000-bit numbers can be multiplied as quickly as 1-bit ones.

Clearly such a model is only an idealization of an actual computer. Able to perform arithmetic in unit time only on words of bounded length, an actual computer cannot efficiently simulate our idealized machine. By starting from 2 and squaring repeatedly, our idealized machine can compute 2^{2^L}—a number represented in binary by a one followed by 2^L 0's—in L time units. A real computer would need time almost 2^L just to write down the answer!

Not only can real computers not compute with large numbers in unit time, they certainly can't compute square roots exactly, given *any* amount of time, since square roots are irrational. A real computer can only approximate a square root.

In theoretical computer science, one usually uses, in place of *our* idealized machine, a random access machine (RAM) whose memory registers can store only integers and for which an operation on b-bit numbers takes time b. Of course exact computation of square roots is forbidden. A RAM would have to approximately compute a square root by, say, Newton iteration. (A RAM can manipulate rationals by representing a rational as a pair of relatively prime integers.) Instead of exact real division, a RAM has truncated division. Such a model more accurately mirrors existing computers.

In the RAM model, the *input size* of a problem instance is the number of bits needed to represent the input, and a problem is said to be solvable in *polynomial time* if there is a program that solves the problem and runs in time at most $p(n)$ for all inputs of size n, for some polynomial $p(n)$. The class of problems solvable in polynomial time is called **P**. Usually considered the class of tractable problems, **P** is extremely important. Any algorithm which solves a problem not in **P** will run extremely slowly on some large instances of the problem.

But how about *our* idealized machine, which, unlike a RAM, performs exact real arithmetic? Despite its theoretical drawbacks, for our purposes exact arithmetic is so convenient that we will use it throughout the book. But before an algorithm can formally be said to be in **P**, one must prove that it can be implemented on the more restricted RAM so as to run in polynomial time. Doing so is usually technically complicated and not very enlightening. All of the algorithms in this book that run in polynomial time on our machine *can* be implemented so as to run in polynomial time on a RAM.

Concerned only with whether our algorithms run in polynomial time, we will make no attempt to present the fastest-known variant of each algorithm. This will help simplify the presentation.

1.3   Linear Algebra and Geometry

Linear algebra plays a central role in the theory of linear programming. As a quick refresher, the most important definitions and theorems we will use are listed below.

Conventions. Throughout this book, all vectors will be column vectors, unless explicitly defined to be row vectors. Sometimes $v_1, v_2, ..., v_n$ will denote the components of n-vector v, sometimes n distinct vectors. Whenever we mean the latter we will say so explicitly. The jth column of an array A (as a column vector) will be denoted A_j; its ith row, *as a row vector*, will be denoted A^i.

A *vector space* or *linear space* S (in $\mathbb{R}^n$) is a nonempty subset of $\mathbb{R}^n$ closed under vector addition and scalar multiplication. Vectors $v_1, \ldots, v_r$ are *linearly independent* if and only if whenever $\sum c_i v_i = 0$ and $c_i \in \mathbb{R}$, then $c_i = 0$ for all i. Vectors that are not linearly independent are *linearly dependent*. The dimension $dim(S)$ of a linear space S is the maximum number of linearly independent vectors in S.

An *affine space* (in $\mathbb{R}^n$) is the translate of a linear space. Formally, $A \subseteq \mathbb{R}^n$ is an affine space if and only if $A = \{\, t + y \mid y \in S \,\}$ for a fixed n-vector t and linear space S. The dimension of A is defined to be $\dim(S)$. By extension, if B is an arbitrary subset of $\mathbb{R}^n$, $\dim(B)$ is defined to be the smallest dimension of any affine space containing B. For example, if $v_1, v_2, \ldots, v_k$ are arbitrary n-vectors, because

$$\{\, v_1, \ldots, v_k \,\} \subseteq \left\{\, v_1 + \sum_{i=2}^{k} \alpha_i(v_i - v_1) \mid \alpha_i \in \mathbb{R} \,\right\},$$

the dimension of $\{\, v_1, \ldots, v_k \,\}$ is at most $k - 1$.

If A is an $m \times n$ matrix, we define

$$column\ space(A) = \{\, Ax \mid x \in \mathbb{R}^n \,\},$$

and

$$rank(A) = \dim(\text{column space}(A)).$$

We say that A has *full rank* if $\mathrm{rank}(A)$ is its smaller dimension. *Nullspace*(A) is the vector space $\{\, x \mid Ax = 0 \,\}$ and the *nullity* of A is the dimension of nullspace(A).

Presented without proof, the first several theorems are standard in the theory of linear algebra.

Theorem 1. Rank(A) is simultaneously the maximum number of linearly independent rows in A and the maximum number of linearly independent columns in A.

Theorem 2. Nullity(A)+ Rank$(A) = n$, if A is $m \times n$.

Where W is a subspace of $\mathbb{R}^n$, let the *orthogonal complement* $W^{\perp}$ of W be $\left\{\, y \in \mathbb{R}^n \mid y^T x = 0 \ \forall x \in W \,\right\}$, the vector space of vectors perpendicular to W.

Theorem 3. Every $x \in \mathbb{R}^n$ can be written uniquely as $x = x^W + x^\perp$, where $x^W \in W$ and $x^\perp \in W^\perp$.

The vector x^W is known as the *projection of x onto W*, while $x^\perp$ is the *projection of x onto $W^\perp$*.

Let A be an $m \times n$ matrix and suppose $At = b$. The set of all solutions to the linear system $Ax = b$ is

$$\{\, x \mid Ax = b \,\} = \{\, t + y \mid Ay = 0 \,\} = \{\, t + y \mid y \in \text{nullspace } (A) \,\};$$

all solutions can be found by adding one particular solution to the nullspace. Using Gaussian Elimination, we can solve a linear system, transform a matrix to upper echelon form, compute its rank and determinant, find its inverse if it's nonsingular, and so on. Gaussian Elimination runs in $O(n^3)$ steps on our exact-arithmetic model. (On a RAM, Gaussian Elimination *can* be implemented so as to run in $O(n^3)$ time—see the chapter-end notes.)

The *length* of a vector x is $\|x\| = \sqrt{x^T x}$.

A set of n-vectors $u_1, u_2, ..., u_m$ is *orthonormal* if each has unit length and distinct vectors have 0 dot product. Alternatively, $u_1, u_2, ..., u_m$ are orthonormal if and only if $u_k^T u_l = 0$ if $k \neq l$, and $u_k^T u_k = 1$ for all k. A real square matrix U is *orthogonal* if $UU^T = I$, i.e., its rows (and hence columns) are orthonormal.

A complex number λ is an *eigenvalue* of A if $Ax = \lambda x$ for some nonzero complex vector x. The vector x is known as the *eigenvector corresponding to λ*.

Theorem 4.   (The **Spectral Theorem**)

If A is a real symmetric matrix, all of A's eigenvalues are real. Furthermore, if A's eigenvalues are $\lambda_1, \lambda_2, ..., \lambda_n$, then

$$A = U \Lambda U^{-1},$$

where

$$\Lambda = \begin{bmatrix} \lambda_1 & 0 & \cdots & 0 \\ 0 & \lambda_2 & & \vdots \\ \vdots & & \ddots & 0 \\ 0 & \cdots & 0 & \lambda_n \end{bmatrix}$$

and U is an orthogonal matrix whose ith column is the eigenvector of A corresponding to λ_i, normalized to have length one.

If A is a real symmetric matrix, we say A is *positive definite* if

$$x^T A x > 0$$

for all nonzero n-vectors x.

Theorem 5. Let A be real symmetric. The following are equivalent.

(a) A is positive definite.

(b) All eigenvalues of A are positive.

(c) $A = QQ^T$ for a nonsingular $n \times n$ real matrix Q.

Now steel yourself for a slew of geometric definitions.

Definition. A *hyperplane* in $\mathbb{R}^n$ is

$$H = \{\, x \mid a_1 x_1 + \cdots + a_n x_n = b \,\},$$

where not all $a_i = 0$. (It is easy to see that $\dim(H) = n - 1$, since if, say, $a_1 \neq 0$, H can be written as $[b/a_1 \ 0 \ 0 \ \cdots \ 0]^T$ plus the nullspace of $1 \times n$ matrix $[a_1 \ a_2 \ \cdots \ a_n]$.) A *halfspace* is the set $\{\, x \mid a_1 x_1 + \cdots + a_n x_n \geq b \,\}$ where not all $a_i = 0$.

We say a set $T \subseteq \mathbb{R}^n$ is *bounded* if there is a real r such that $\|x\| \leq r$ for all $x \in T$.

Definition. A *polyhedron* is the intersection of finitely many halfspaces. A bounded nonempty polyhedron is called a *polytope*.

Example. The set $\{\, x \mid Ax \geq b, \ x \geq 0 \,\}$ is the intersection of finitely many halfspaces, as is $\{\, x \mid Ax = b, \ x \geq 0 \,\}$. If they are nonempty and bounded, they're polytopes.

Example. Here is a polytope in $\mathbb{R}^2$:

$$
\begin{aligned}
2x_1 + x_2 &\geq 4 \\
x_1 - x_2 &\geq -4 \\
-3x_1 + x_2 &\geq -15 \\
-x_1 &\geq -7 \\
x_1 \geq 0 \ x_2 &\geq 0
\end{aligned}
$$

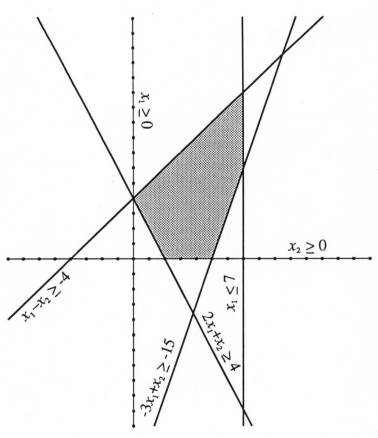

We say a sequence of equality or inequality constraints is *linearly independent* if the sequence of coefficient vectors is linearly independent. The right-hand sides and whether the constraints are equalities or inequalities are ignored.

Definition. If $x, y, p \in \mathbb{R}^n$ and $p = \lambda x + (1 - \lambda)y$, $0 \leq \lambda \leq 1$, then p is the *convex combination* of x and y. More generally, p is the *convex combination* of points $x_1, x_2, \ldots, x_r$ if $p = \sum_{i=1}^{r} \lambda_i x_i$, $\lambda_i \geq 0$, and $\sum_{i=1}^{r} \lambda_i = 1$. Point p is the *strict convex combination* of points $x_1, x_2, \ldots, x_r$ if p is the convex combination of $x_1, x_2, \ldots, x_r$ yet $p \neq x_i$ for all i.

Definition. $S \subseteq \mathbb{R}^n$ is *convex* if whenever $x, y \in S$, $\lambda x + (1 - \lambda)y \in S$ for all $\lambda \in [0,1]$. That is, the line segment between x and y lies in S.

It is easy to see that (a) the intersection of arbitrarily many convex sets is convex, and (b) a convex combination of points in a convex set is itself in the set.

1.4 Basic Solutions

Definition. If P is a polyhedron, $v \in P$ is an *extreme point* or *vertex* if v cannot be written as the strict convex combination of points $x, y \in P$.

If v is extreme, one can't "slide" along any line through v and stay within P. Not even a 1-dimensional ball around v lies entirely in P.

Definition. If A is an $m \times n$ matrix of rank m, any m linearly independent columns are a *basis*. (They form a basis of A's column space, $\mathbb{R}^m$). Sometimes we will use the columns as the basis, sometimes just their indices.

Definition. Consider the linear program $\min c^T x$ where A is of

$$Ax = b$$
$$x \geq 0$$

rank m and is $m \times n$. If columns $j_1 < j_2 < \cdots < j_m$ are a basis, the corresponding *basic solution* x is defined as follows and ignores c. Let B be the nonsingular matrix consisting of columns $j_1, j_2, \ldots,$ j_m of A. Any variable x_l where $l \notin \{ j_1, j_2, \ldots, j_m \}$ is set to 0 and is called *nonbasic*. For $k = 1, \ldots, m$, x_{j_k} is set to the kth component of $B^{-1}b$ and is known as *basic*; the basic variables may or may not be 0. (The basic solution x is the unique vector satisfying $Ax = b$ subject to the condition that $x_l = 0$ for $l \notin \{ j_1, \ldots, j_m \}$.) If x satisfies the sign constraints $x \geq 0$, x is a *basic feasible solution* or *bfs*.

$$\boxed{\text{BASIC SOLUTIONS ARE VERY IMPORTANT!}}$$

Example.

$$
\begin{aligned}
x_1 + 2x_2 - 3x_3 + x_4 &= 7 \\
2x_1 + 0x_2 - 7x_3 + x_4 &= 9 \\
x_j &\geq 0
\end{aligned}
$$

One basis is $\{ 3, 4 \}$ or $\left\{ \begin{bmatrix} -3 \\ -7 \end{bmatrix}, \begin{bmatrix} 1 \\ 1 \end{bmatrix} \right\}$. Because

$$
\begin{bmatrix} -3 & 1 \\ -7 & 1 \end{bmatrix}^{-1} = \begin{bmatrix} \frac{1}{4} & -\frac{1}{4} \\ \frac{7}{4} & -\frac{3}{4} \end{bmatrix},
$$

the basic solution corresponding to basis $\{ 3, 4 \}$ is the vector $[0 \;\; 0 \;\; -\frac{1}{2} \;\; \frac{11}{2}]^T$. As $x_3 < 0$, it is infeasible.

If $m \times n$ matrix A has rank m, then $Ax = b$ always has a basic $x \geq 0$ solution. If the feasible region is nonempty, we will see that it must have a bfs as well.

Notation. Let us use *LPS* to denote

" $\min c^T x$,   where A is an $m \times n$ matrix of rank m."
s.t. $Ax = b$
    $x \geq 0$

We use F to denote the feasible set $\{\,x \mid Ax = b,\ x \geq 0\,\}$. Throughout, i will generally denote a row index, running from 1 to m, and j will generally denote a column index, running from 1 to n.

Theorem 6. In LPS, suppose that $w \in \mathbb{R}^n$. Then w is a vertex of the feasible set F if and only if w is a basic feasible solution of the LPS instance.

Proof. If w is a vertex of F, w is feasible. Let $I = \{\,i \mid w_i > 0\,\}$. Recall that A_j denotes the jth column of A.

First we prove that w is a bfs if $\{\,A_j \mid j \in I\,\}$ is independent. Since A's rank is m and $|I| \leq m$, we can extend I to an m-element set $J \supseteq I$ such that $\{\,A_j \mid j \in J\,\}$ is independent. J is a basis and $w_j = 0$ if $j \neq J$. Therefore w is the bfs corresponding to J.

So suppose $\{\,A_j \mid j \in I\,\}$ is dependent. Then there is a nonzero z satisfying $\sum_{j \in I} z_j A_j = 0$. Define $z_j = 0$ if $j \notin I$, so that $Az = 0$. For $j \in I$, $w_j > 0$. Let

$$\theta = \min_{z_j \neq 0} \frac{w_j}{|z_j|}.$$

Clearly $\theta > 0$.

Now let $w^+ = w + \theta z$, $w' = w - \theta z$. $Aw^{\pm} = Aw \pm \theta(Az) = b$. For $j \notin I$, $w_j^{\pm} = w_j \geq 0$. Thus $\theta|z_j| \leq \dfrac{w_j}{|z_j|} \cdot |z_j| = w_j$. Hence $w_j \pm \theta z_j \geq 0$ if $j \in I$ and thus for all j, $w_j^{\pm} \geq 0$. So w^+ and $w^- \in F$ (since $Aw^{\pm} = b$), yet since $\theta > 0$ and $z \neq 0$, neither w^+ nor w^- is w. Yet $w = \frac{1}{2}w^+ + \frac{1}{2}w^-$, a contradiction.

Now let w be a bfs of $Ax = b$, $x \geq 0$, where A is of rank m. Suppose that $w = \lambda u + (1 - \lambda)v$, where $0 < \lambda < 1$ and $u, v \in F$, $u \neq w, v \neq w$. Choose a basis $J \subseteq \{\,1, 2, \ldots, n\,\}$ corresponding to w. For $j \notin J$, $w_j = 0$. Furthermore, w is the unique vector in F satisfying $w_j = 0$ if $j \notin J$. Thus there is a $j \notin J$ such that $u_j > 0$. Because $v \geq 0$, $v_j \geq 0$. Thus $\lambda u_j + (1-\lambda)v_j$, being the jth coordinate of $\lambda u + (1 - \lambda)v$, is strictly positive, and therefore $w \neq \lambda u + (1 - \lambda)v$, a contradiction. ∎

Corollary 7. There are only finitely many vertices.

Proof. Each is a bfs, which is determined by an m-element subset of the columns. The number of bfs's is therefore at most $\binom{n}{m}$. ∎

Definition. A bfs is *degenerate* if it has more than $n - m$ zeroes.

Example. In the linear system

$$x_1 + x_2 - 3x_3 = 5$$
$$2x_1 - x_2 + 3x_3 = 10$$

$$x_1 \geq 0 \ x_2 \geq 0 \ x_3 \geq 0,$$

$$A = \begin{bmatrix} 1 & 1 & -3 \\ 2 & -1 & 3 \end{bmatrix},$$

$n = 3$, $m = 2$, and $n - m = 1$.

Corresponding to basis $\{1, 2\}$ is bfs $\begin{bmatrix} 5 \\ 0 \\ 0 \end{bmatrix}$, which has more than

$n - m$ zeroes. Bfs $\begin{bmatrix} 5 \\ 0 \\ 0 \end{bmatrix}$ also corresponds to basis $\{1, 3\}$.

Theorem 8. If two different bases correspond to a single bfs v, then v is degenerate.

(The converse is false.)

Proof. Let J_1 and J_2 be the two bases corresponding to v. If $j \notin J_1 \cap J_2$ then v_j must be 0. But $|J_1 \cap J_2| < m$, so v must have more than $n - m$ zeroes. ∎

A constraint is *tight* if it is satisfied with equality.

Lemma 9. Suppose $P = \{x \in \mathbb{R}^n | Ax \leq b\}$ is a polyhedron. Then v is an extreme point of P if and only if there are n linearly independent constraints among the constraints $Ax \leq b$ that are tight at v.

Proof. $\Rightarrow$: Let v be an extreme point. Let $Bx \leq b'$ be those constraints among the constraints $Ax \leq b$ that are tight at v (so $Bv = b'$), and suppose for a contradiction that $\text{rank}(B) \leq n - 1$. Choose a nonzero vector t in the nullspace of B. Then

$$v' = v + \gamma t$$

satisfies $Bv' = b'$ for any real γ, and for a sufficiently small positive γ_0 both $v + \gamma_0 t$ and $v - \gamma_0 t$ satisfy the remaining constraints. But

$$v = \frac{1}{2}(v + \gamma_0 t) + \frac{1}{2}(v - \gamma_0 t),$$

contradicting the fact that v is extreme.

$\Leftarrow$: Suppose that the constraints $Bx \leq b'$ from among the constraints $Ax \leq b$ are tight at v and $\text{rank}(B) = n$. Suppose for a contradiction that

$$v = \lambda v_1 + (1 - \lambda) v_2,$$

$0 < \lambda < 1$, and $v_1 \neq v$, $v_2 \neq v$, yet $v_1, v_2 \in P$. Because $Bv_1 \leq b'$ and $Bv_2 \leq b'$ and $0 < \lambda < 1$,

$$Bv = \lambda(Bv_1) + (1 - \lambda)(Bv_2) = b'$$

if and only if $Bv_1 = Bv_2 = b'$. But because B has rank n, it follows that $v_1 = v_2 = v$, a contradiction. ∎

Since an equality is a pair of inequalities, the same result holds even if P is defined by both inequalities and equalities.

Say a minimization problem is *unbounded* if for each real B, there is a feasible point of cost less than B. We define unbounded maximization problems similarly.

Theorem 10. In LPS, suppose that $p \in F$. Then either the LPS instance is unbounded or there is a vertex v of F satisfying $c^T v \leq c^T p$.

Here is a corollary.

Corollary 11. If in LPS $F \neq \emptyset$ and $c^T x \geq B$ for all $x \in F$, then there is an optimal vertex. Provided that the cost is bounded below, linear programming is a finite problem!

Proof sketch. Let $p_1 \in F$. Find a vertex v_1 such that $c^T v_1 \leq c^T p_1$. (Theorem 10 guarantees v_1's existence.) If v_1 is optimal we are done. Otherwise, there is a feasible point p_2 such that $c^T p_2 < c^T v_1$, and a vertex v_2 such that $c^T v_2 \leq c^T p_2 < c^T v_1$. If v_2 is optimal we are done. Otherwise, there is a feasible point p_3 such that $c^T p_3 < c^T v_2$, and a vertex v_3 such that $c^T v_3 \leq c^T p_3 \leq c^T v_2 < c^T v_1$. If v_3 is optimal we are done. In this way we construct a sequence of vertices $v_1, v_2, v_3, \ldots$ each of cost strictly less than the previous one's. No vertex can appear twice on this list, and since there are only finitely many vertices, the process must terminate with an optimal bfs. ∎

Proof of Theorem 10. Because p is feasible, $Ap = b$. Number the constraints from 1 to $m + n$, the n sign constraints first. Among all sets of linearly independent constraints that contain the constraints $Ax = b$ *all of which are tight at p,* choose a largest one $S \subseteq \{1, 2, \ldots, m + n\}$. Let $r = |S|$ and let $I = S \cap \{1, \ldots, n\}$, the $r - m$ sign constraints in S. If $r = n$, from Lemma 9 we infer that p is a vertex and there is nothing to prove. If $r < n$, then we show that either $c^T x$ for $x \in F$ can be made arbitrarily small, or that there is another point $p^* \in F$ of cost no more than the cost of p, for which there is a larger set of linearly independent constraints tight at p^* (including the $Ax = b$ constraints). Iterating this process at most $n - m$ times, we either find a proof that the cost is unbounded, or we reach a vertex v such that $c^T v \leq c^T p$.

Observation. Since we chose S to be as large as possible, we know that if $x_j \geq 0$ is a sign constraint tight at p and $j \notin S$, then this constraint must be dependent on the tight constraints S we selected.

Let

$$F^* = \{\, x \mid \underbrace{x_j = 0 \ \forall j \in I, \text{ and } Ax = 0}_{\text{constraints in } S \text{ with RHS's set to } 0} \,\},$$

the nullspace of an $r \times n$ matrix of rank r. ("RHS" denotes "right-hand side.") $\text{Dim}(F^*) = n - r > 0$. Use Gaussian Elimination to find a nonzero vector w in F^*. We may assume $c^T w \leq 0$ by negating w if necessary.

Claim. If $p_j = 0$ then $w_j = 0$.

Proof of Claim. By the definition of F^*, $w_j = 0$ for $j \in I$.

If $p_j = 0$ and $j \notin I$, then by the observation, the constraint $x_j \geq 0$, tight p, must be linearly dependent on the constraints in S. So since a linear combination of the 0 RHS's is 0, w_j also equals 0. The claim follows.

Now for all λ, $A(p + \lambda w) = Ap + \lambda \underbrace{Aw}_{0} = b$.

If $p_j = 0$, $(p + \lambda w)_j = p_j + \lambda w_j = 0$, by the claim.

So for all real λ, $p + \lambda w$ satisfies the equality constraints and $p_j = 0$ implies $(p + \lambda w)_j = 0$. We will prove that either one can choose λ's arbitrarily large in such a way that $p + \lambda w$ stays feasible and $c^T(p + \lambda w)$ can be made arbitrarily small, or there is a specific value for λ such that $c^T(p + \lambda w) \leq c^T p$ and at least one more sign constraint is tight at $p + \lambda w$ than at p (and that sign constraint is linearly independent from the constraints in S).

Let $J = \{ j \mid w_j < 0 \}$. We have three cases.

Case 1. $c^T w < 0$ and $J = \emptyset$.

Since $J = \emptyset$, $w_j \geq 0$ for all j. So $p + \lambda w \in F$ for all $\lambda \geq 0$. But

$$c^T(p + \lambda w) = c^T p + \lambda \underbrace{(c^T w)}_{<0},$$

so the objective function can be made arbitrarily small while maintaining feasibility.

Case 2. $J \neq \emptyset$.

Let $s = \min_{j \in J} \dfrac{p_j}{-w_j}$. Since $j \in J$ implies that $p_j > 0$ and $w_j < 0$, it follows that $s > 0$. Choose $j^* \in J$ such that $s = \dfrac{p_{j^*}}{-w_{j^*}}$ and let $p^* = p + sw$. For all $j \in J$,

$$p_j^* = p_j + sw_j \geq p_j + \frac{p_j}{-w_j} w_j = p_j - p_j = 0.$$

For $j \notin J$, $p_j^* \geq 0$ also. And

$$p_{j^*}^* = p_{j^*} + \frac{p_{j^*}}{-w_{j^*}} \cdot w_{j^*} = 0,$$

whereas $p_{j^*} > 0$—one more sign constraint is tight at p^* than at p. (Both p^* and p satisfy all the constraints in S with equality. If the constraint $x_{j^*} \geq 0$ were *dependent* on S, then it would be impossible for $p_{j^*}^* \neq p_{j^*}$.) Now

$$\begin{aligned} Ap^* &= Ap + \lambda Aw = b, \\ c^T p^* &= c^T p + s(c^T w) \leq c^T p. \end{aligned}$$

Case 3. $c^T w = 0$ and $J = \emptyset$.

For all j, $w_j \geq 0$. Let

$$s = \underbrace{\min_{w_j > 0}}_{\text{nonempty}} \frac{p_j}{w_j};$$

$s > 0$. Choose j^* such that $s = \dfrac{p_{j^*}}{w_{j^*}}$ (and $p_{j^*}, w_{j^*} > 0$), and let $p^* = p - sw$. Note that $p_{j^*}^* = 0$ and clearly $Ap^* = Ap = b$.

Suppose j satisfies $p_j > 0$ and $w_j > 0$. Then $p_j^* = p_j - sw_j$. Since $s \leq \dfrac{p_j}{w_j}$, $-sw_j \geq -p_j$. Thus $p_j^* \geq p_j - p_j = 0$.

Otherwise $w_j = 0$ and $p_j^* = p_j \geq 0$. It follows that p^* is feasible. But $p_{j^*}^* = 0$ while $p_{j^*} > 0$. So p^* satisfies at least one more sign constraint with equality than did p, and as before, this constraint is linearly independent from the constraints in S. Last,

$$c^T p^* = c^T p - s \underbrace{(c^T w)}_{0} = c^T p. \quad \blacksquare$$

Note that this proof yields a polynomial-time algorithm to find a vertex v such that $c^T v \leq c^T p$. This technique can also be used to show that in LPS, if $F \neq \emptyset$ then some bfs exists (even if the problem is unbounded), and that every polytope has a vertex.

Theorem 12. If $P = \{x \in \mathbb{R}^n | Ax \leq b\}$ is a polytope, then every $x \in P$ can be written as a convex combination of at most $n + 1$ vertices.

Proof. Fix n. We will prove, by downward induction on r, that if $P \subseteq \mathbb{R}^n$ is a polytope, x is in P and there are r or more linearly independent constraints which are tight at x, then x can be written as a convex combination of at most $n + 1 - r$ vertices of P.

Basis. If $r = n$, by Lemma 9 x is already a vertex.

Inductive Step. Choose $r < n$. Let $P = \{x \in \mathbb{R}^n | Ax \leq b\}$ be a polytope and let $x \in P$ be satisfied with equality by at least r linearly independent constraints. We may assume x is not a vertex. Build a new polytope by replacing all tight inequalities with equalities. Since every vertex of the new polytope is a vertex of P, we may replace P by the new polytope, also called P.

Since P is a polytope (not just a polyhedron), P has a vertex, say v. Consider $v + t(x - v)$ for $t \geq 1$. Let $t_0 = \max\{t | v + t(x - v) \in P\} > 1$. At $y = v + t_0(x - v)$, at least one more constraint is tight than at x. Construct P' from P by changing that one constraint to an equality. Since at least $r + 1$ linearly independent constraints are tight at y, by the inductive hypothesis we can write y as a convex combination of at most $n + 1 - (r + 1)$ vertices of P' (and of P):

$$y = \sum_{l=1}^{n-r} \lambda_l v_l,$$

$\sum \lambda_l = 1$, $\lambda_l \geq 0$ for all l. Now $y = v + t_0(x - v)$ implies that $(1/t_0)y + (1 - 1/t_0)v = x$. Therefore

$$\sum_{l=1}^{n-r} \frac{\lambda_l}{t_0} v_l + (1 - \frac{1}{t_0})v = x;$$

x is a convex combination of at most $n + 1 - r$ vertices. ∎

1.5   Notes

Written with computational issues in mind, Strang's book on linear algebra [56] not only covers the relevant theorems, but it also shows how one can algorithmically find inverses, eigenvalues, etc. It even includes a chapter on linear programming. The text by Hoffman and Kunze [31] has a more theoretical perspective. A classic text in abstract algebra, Herstein's text [29] covers everything from groups to rings to fields, in addition to linear algebra.

The RAM model is discussed carefully in Aho, Hopcroft and Ullman's *The Design and Analysis of Computer Algorithms* [6]. A RAM, unlike the idealized computer we've used, can perform arithmetic operations in constant time only on words of limited length, and formally the distinction is crucial. Our idealized machine is so much simpler conceptually that we've adopted it despite its drawbacks. The reader can find a proof that Gaussian Elimination can be implemented so as to run in $O(n^3)$ time even on a traditional RAM in either [27] or [47]. Edmonds' original proof that Gaussian Elimination can be made to run in polynomial time appears in [18].

Any book on the theory of algorithms, such as those by Aho, Hopcroft and Ullman [6], Brassard and Bratley [14], Baase [9] and Manber [41], covers $O()$ and related notation.

The theorem alluded to on the integrality of solutions to linear programs for bipartite matching problems can be found in the section on total unimodularity in the text by Papadimitriou and Steiglitz [44]. There, it is proven that if every square submatrix of A has determinant 0, 1, or -1, then every linear program in general form with constraint matrix A has an integral optimal point, if it has any at all; and that the constraint matrices for bipartite matching problems have this property. Total unimodularity is also discussed in [39].

Chapter 2

The Simplex Algorithm

Designed in 1947 by G. Dantzig, the *Simplex Algorithm* was the method of choice used to solve linear programs for decades. Though not a polynomial-time algorithm in the worst case, the Simplex Algorithm is remarkably fast in practice. Problems with thousands of variables and constraints are routinely solved by the Simplex Algorithm. Only since the appearance of Karmarkar's Algorithm in 1984 and the more recent interior-point methods have contenders for "best practical linear programming algorithm" existed.

Conceptually, the Simplex Algorithm could hardly be simpler. We already know that linear programming is a finite problem. When the objective function is bounded below, we need only scan all of the (exponentially many) vertices of the polytope in order to find an optimal point. The Simplex Algorithm is nothing more than an orderly way of scanning the vertices.

But not all of them! That would take too long. The algorithm starts from an arbitrary vertex v of the feasible region and (ideally) finds a cheaper neighboring vertex, a neighboring vertex being a different bfs whose corresponding basis differs from that of v in exactly one column. If no neighboring vertex has cheaper cost, the current vertex can be proven globally optimal. What could be simpler? There are some complications caused by degeneracy and the possibility that the problem is unbounded, but they're not too difficult to handle.

2.1 Pivoting

We start with an example.

Example.

$$\text{Min } 2x_2 + x_4 + 5x_7$$
$$\begin{aligned}
\text{s.t. } 2x_1 + x_2 + 2x_3 + 2x_4 + x_5 &= 16 \\
x_1 + 4x_3 + 2x_5 &= 16 \\
-x_1 + 2x_2 + x_3 - 5x_4 + 5x_5 + 2x_6 &= 34 \\
3x_2 - x_3 + x_4 + 4x_5 + 4x_6 &= 38 \\
x_i &\geq 0 \ \forall i
\end{aligned}$$

$A = \begin{bmatrix} 2 & 1 & 2 & 2 & 1 & 0 \\ 1 & 0 & 4 & 0 & 2 & 0 \\ -1 & 2 & 1 & -5 & 5 & 2 \\ 0 & 3 & -1 & 1 & 4 & 4 \end{bmatrix}$, $b = \begin{bmatrix} 16 \\ 16 \\ 34 \\ 38 \end{bmatrix}$. You magically learn

that $\{1, 2, 5, 6\}$ is a basis with bfs $\begin{bmatrix} 4 \\ 2 \\ 0 \\ 0 \\ 6 \\ 2 \end{bmatrix}$; therefore $4A_1 + 2A_2 +$

$6A_5 + 2A_6 = b = \begin{bmatrix} 16 \\ 16 \\ 34 \\ 38 \end{bmatrix}$.

Nonbasic column $A_4 = \begin{bmatrix} 2 \\ 0 \\ -5 \\ 1 \end{bmatrix}$ can be written as this linear combination of the basic columns A_1, A_2, A_5, A_6:

$$A_4 = 2 \cdot A_1 - 1 \cdot A_2 - 1 \cdot A_5 + 2 \cdot A_6.$$

Let's consider increasing the value of nonbasic x_4 from 0 to some value θ, while maintaining feasibility and leaving nonbasic x_3 at 0. Let x'_1, x'_2, x'_5 and x'_6 be the new values, respectively, of x_1, x_2, x_5 and x_6. We want $(x'_1 A_1 + x'_2 A_2 + x'_5 A_5 + x'_6 A_6) + \theta A_4 = \begin{bmatrix} 16 \\ 16 \\ 34 \\ 38 \end{bmatrix}$. From

$A_4 = 2A_1 - A_2 - A_5 + 2A_6$, this is equivalent to

$$(x'_1 + 2\theta)A_1 + (x'_2 - \theta)A_2 + (x'_5 - \theta)A_5 + (x'_6 + 2\theta)A_6 = \begin{bmatrix} 16 \\ 16 \\ 34 \\ 38 \end{bmatrix}.$$

This is clearly satisfied if $x'_1 + 2\theta = x_1$, $x'_2 - \theta = x_2$, $x'_5 - \theta = x_5$, and $x'_6 + 2\theta = x_6$, in other words, $x'_1 = 4 - 2\theta$, $x'_2 = 2 + \theta$, $x'_5 = 6 + \theta$, and $x'_6 = 2 - 2\theta$. Any θ will satisfy the equality constraints, but to remain feasible, we must worry about the sign constraints. Therefore we need:

$$\begin{array}{cccc} 4 - 2\theta \geq 0 & 2 + \theta \geq 0 & 6 + \theta \geq 0 & 2 - 2\theta \geq 0, \\ \text{i.e.,} \quad \theta \leq 2 & \theta \geq -2 & \theta \geq -6 & \theta \leq 1. \end{array}$$

Let us take $\theta = 1$, the largest value of θ that will maintain feasibility. The new point x' is $[2\ 3\ 0\ 1\ 7\ 0]^T$ and is a bfs; column 4 enters the basis and column 6 leaves. We will see that by increasing a nonbasic variable as much as possible (and "bumping into" a sign constraint), we reach a new bfs. This process is called a *pivot*.

Let's do another example of a pivot. Suppose that in LPS

$$A = \begin{bmatrix} 5 & 2 & -3 & 16 & 4 \\ 2 & 3 & 1 & 3 & 1 \\ 1 & 7 & 6 & -1 & 2 \end{bmatrix}, \quad b = \begin{bmatrix} 8 \\ 8 \\ 25 \end{bmatrix}$$

and our current bfs is

$$x_0 = \begin{bmatrix} 1 \\ 0 \\ 3 \\ 0 \\ 3 \end{bmatrix}$$

with corresponding basis $\{1, 3, 5\}$. Preferred by mechanical pivoting methods, an *ordered basis* is a basis with its elements listed in some specified order. Let us order the columns, say, $5, 1, 3$, define $B(1) = 5$, $B(2) = 1$, $B(3) = 3$, and view our basis as an ordered basis $\mathcal{B} = B(1), B(2), B(3)$. Where x_0 is the current bfs, let x_{i0} denote the $B(i)$th entry of x_0, $i = 1, 2, ..., m$; x_0's other entries are 0.

Let us attempt to bring nonbasic column $j = 4$ into the basis. Because the basic columns span the column space, every nonbasic column (indeed, *every* m-vector) can be written as a linear combination of the basis vectors. In our case,

$$A_4 = 2A_5 + 1A_1 + (-1)A_3.$$

We will use x_{ij} to denote the coefficient of $A_{B(i)}$ in the representation of nonbasic column j as a linear combination of the basic columns. Thus $x_{14} = 2$, $x_{24} = 1$, and $x_{34} = -1$, so that

$$A_j = \sum_{i=1}^{m} x_{ij} A_{B(i)} = x_{14} A_{B(1)} + x_{24} A_{B(2)} + x_{34} A_{B(3)}.$$

To pivot we must increase the jth component of x_0, and change components $B(1), B(2), ..., B(m)$ of x_0 so as to maintain feasibility. How? Because

$$A_4 = 2A_5 + 1A_1 + (-1)A_3,$$

we see that if we increase the jth component of x_0 by θ, to counterbalance this increase, the fifth entry of x_0, now 3, should decrease by 2θ, the first entry of x_0, now 1, should decrease by 1θ, and the third entry, now 3, should decrease by -1θ (i.e., increase by θ). To maintain nonnegativity, we thus need $3 - 2\theta$, $1 - 1\theta$, and $3 - (-1)\theta$ all to be nonnegative, and therefore θ cannot exceed $\theta_0 = \min\{\frac{3}{2}, \frac{1}{1}, -\} = 1$. (Ignore the ratio with negative denominator!)

Now we compute the new values of x_4, x_5, x_1, x_3 :

1. $x_4 = \theta_0 = 1$. Column 4 enters the basis.

2. $x_5 = 3 - 2\theta_0 = 1$.

3. $x_1 = 1 - \theta_0 = 0$. Column 1 leaves the basis.

4. $x_3 = 3 + \theta_0 = 4$.

The new ordered basis is $B(1) = 5$, $B(2) = 4$, $B(3) = 3$ and the new bfs is $[0 \ \ 0 \ \ 4 \ \ 1 \ \ 1]^T$.

Formalizing this procedure, let x_0 be a bfs corresponding to ordered basis $\mathcal{B} = B(1), B(2), \ldots, B(m)$. Let x_{i0} denote the $B(i)$th component of x_0, $1 \leq i \leq m$. Then $Ax_0 = b$ implies $\sum_{i=1}^{m} x_{i0} A_{B(i)} = b$. Because $\mathcal{B}$ is a basis, we can write each nonbasic column j as a linear combination of the basic columns for certain reals x_{ij}:

$$A_j = \sum_{i=1}^{m} x_{ij} A_{B(i)}.$$

Contemplate increasing the value of the jth entry of x_0 from 0 to θ. Since $\theta A_j = \sum_{i=1}^{m} x_{ij} \theta A_{B(i)}$,

$$\sum_{i=1}^{m} x_{ij} \theta A_{B(i)} - \theta A_j = 0$$

and therefore

$$\sum_{i=1}^{m} x_{i0} A_{B(i)} - \sum_{i=1}^{m} x_{ij} \theta A_{B(i)} + \theta A_j = b.$$

In other words,

$$\sum_{i=1}^{m} (x_{i0} - x_{ij}\theta) A_{B(i)} + \theta A_j = b. \tag{2.1}$$

Let us build an n-vector x_0' in this way:

- Its jth component is θ.

- For each i, its $B(i)$th component is $x_{i0} - x_{ij}\theta$.

- Its other components are 0.

Then (2.1) shows that x_0' satisfies the equality constraints $Ax = b$. If in addition $x_{i0} - x_{ij}\theta \geq 0$ for $i = 1, \ldots, m$, then x_0' also satisfies the sign constraints. We will always choose

$$\theta = \theta_0 = \min_{i:x_{ij}>0} \frac{x_{i0}}{x_{ij}},$$

the largest value for θ that will maintain feasibility.

Potential Problems:

1. There might not be any i such that $x_{ij} > 0$. We can then choose any $\theta_0 \geq 0$ and remain feasible.

2. θ_0 may be 0, in which case $x_0' = x_0$ (but the ordered bases differ). A pivot with $\theta_0 = 0$ is called *degenerate*.

Notation. The bfs x_0 has n entries, of which at least $n - m$ are 0. Where x_{i0} denotes the $B(i)$th entry of x_0, sometimes we will represent x_0 by the m-vector $[x_{10} \ x_{20} \ \cdots \ x_{m0}]^T$ of basic entries.

We will show that if the minimum in the definition of θ_0 is taken over a nonempty set, then x_0' is a bfs. In fact, if the minimum in the definition of θ_0 is achieved by $\dfrac{x_{l0}}{x_{lj}} = \theta_0$ (where $x_{lj} > 0$) then a basis corresponding to x_0' is given by

$$B'(i) = \begin{cases} B(i) & \text{if } i \neq l \\ j & \text{if } i = l. \end{cases}$$

Column j replaces column $B(l)$ in the ordered basis.

Example. $A = \begin{bmatrix} 1 & 2 & 2 & 1 & 0 & 1 \\ 1 & 1 & 3 & 3 & 1 & 0 \end{bmatrix}, b = \begin{bmatrix} 5 \\ 3 \end{bmatrix}.$

Let $B(1) = 6$, $B(2) = 5$. Then $x_0 = \begin{bmatrix} 0 \\ 0 \\ 0 \\ 0 \\ 3 \\ 5 \end{bmatrix}$ and $x_{10} = 5$, $x_{20} = 3$.

Let $j = 3$. A_j is nonbasic and equals $2A_6 + 3A_5$. (Because the basic columns form an identity matrix, finding the coefficients is easy.) Then $x_{13} = 2$, $x_{23} = 3$ and

$$\theta_0 = \min_{i:x_{ij}>0} \frac{x_{i0}}{x_{ij}} = \min\left\{\frac{5}{2}, \frac{3}{3}\right\} = 1 = \frac{x_{20}}{x_{2j}};$$

$l = 2$. The new ordered basis is

$$B'(i) = \begin{cases} B(i) & \text{if } i \neq 2 \\ 3 & \text{if } i = 2, \end{cases}$$

i.e., $B'(1) = 6$, $B'(2) = 3$. The new bfs is $\begin{bmatrix} 0 \\ 0 \\ \theta_0 \\ 0 \\ 3 - \theta_0 \cdot 3 \\ 5 - \theta_0 \cdot 2 \end{bmatrix} = \begin{bmatrix} 0 \\ 0 \\ 1 \\ 0 \\ 0 \\ 3 \end{bmatrix}$.

Theorem 13. Given ordered basis $\mathcal{B} = B(1), B(2), \ldots, B(m)$ and the bfs x_0 corresponding to $\mathcal{B}$ (whose $B(i)$th component is x_{i0}, $1 \leq i \leq m$), let $j \notin \mathcal{B}$. Write $A_j = \sum_{i=1}^{m} x_{ij} A_{B(i)}$. Suppose $\{i \mid x_{ij} > 0\}$ is nonempty. Let θ_0 denote $\min_{i:x_{ij}>0} \frac{x_{i0}}{x_{ij}}$ and suppose $\theta_0 = \frac{x_{l0}}{x_{lj}}$ (with $x_{lj} > 0$). Let $B'(i) = \begin{cases} B(i), & i \neq l \\ j, & i = l \end{cases}$. Then $\mathcal{B}' = B'(1), B'(2), \ldots, B'(m)$ is a basis corresponding to the bfs x_0' whose $B'(i)$th component is $x_{i0} - \theta_0 x_{ij}$, if $i \neq l$, and whose jth component is θ_0.

Proof. By the discussion above, and the fact that the tth entry of x_0' is 0 if $t \notin \mathcal{B}'$, x_0' is feasible. We need prove only that columns $A_{B'(1)}, A_{B'(2)}, \ldots, A_{B'(m)}$ are linearly independent.

For $i \neq l$, $A_{B(i)} = A_{B'(i)}$.

$$A_j = \left(\sum_{i \neq l} x_{ij} A_{B(i)} \right) + A_{B(l)} x_{lj}$$

$$A_j = A_{B'(l)} = \left(\sum_{i \neq l} x_{ij} A_{B'(i)} \right) + A_{B(l)} x_{lj},$$

so

$$A_{B(l)} = \frac{A_{B'(l)} - \sum_{i \neq l} x_{ij} A_{B'(i)}}{x_{lj}}.$$

The denominator is nonzero. Therefore the m column vectors $A_{B'(1)}, A_{B'(2)}, \ldots, A_{B'(m)}$ span the rank-m set $A_{B(1)}, A_{B(2)}, \ldots, A_{B(m)}$ and thus must be linearly independent. ■

Note. If there is a tie in the min operation that determines θ_0, the new bfs is degenerate—we get at least two 0's "at the cost of one."

2.2 Tableaux

The idea here is that if the columns $A_{B(1)}, \ldots, A_{B(m)}$ constituted an identity matrix, pivoting would be easy. In that case, we could write a nonbasic column A_j as $\sum_{i=1}^{m} x_{ij} A_{B(i)}$ trivially:

$$A_j = \sum_{i=1}^{m} a_{ij} A_{B(i)}$$

(where a_{ij} is the (i,j) entry of A). We will force the basic columns to be an identity matrix by executing elementary row operations.

Notation. For $1 \leq i \leq m$, let e_i be the ith standard unit m-vector, the m-vector whose ith entry is one but which is otherwise 0. Clearly

$$I = \begin{bmatrix} | & | & & | \\ e_1 & e_2 & \cdots & e_m \\ | & | & & | \end{bmatrix}.$$

We want $A_{B(i)} = e_i$. Suppose we have the system

$$
\begin{aligned}
6 &= x_1 + x_2 + 4x_6 \\
14 &= -2x_1 + 2x_2 + x_3 - x_4 + x_6 \\
-11 &= x_1 - 2x_2 + x_4 + 2x_6 \\
7 &= x_1 - 3x_4 + x_5 - 5x_6 \\
x_j &\geq 0 \ \forall j.
\end{aligned}
$$

Represent it as an $m \times (n + 1)$ matrix, a *tableau*:

	x_1	x_2	x_3	x_4	x_5	x_6
6	1	1	0	0	0	4
14	-2	2	1	-1	0	1
-11	1	-2	0	1	0	2
7	1	0	0	-3	1	-5

The leftmost column is known as column 0.

Suppose we choose $B(1) = 2$, $B(2) = 3$, $B(3) = 4$, $B(4) = 5$. Though these four columns are linearly independent, they do not contain an identity matrix. Without changing the solution space, we can put e_i in $A_{B(i)}$ for each i by executing elementary row operations (on all $n + 1$ columns), obtaining

	x_1	x_2	x_3	x_4	x_5	x_6
6	1	1	0	0	0	4
3	-1	0	1	0	0	3
1	3	0	0	1	0	10
10	10	0	0	0	1	25

By virtue of having the identity matrix in columns $A_{B(1)}$, $A_{B(2)}$, $\ldots$, $A_{B(m)}$, the ith entry in column 0 gives the $B(i)$th component in the current bfs! That is, if X is the name of this matrix, the ith entry in X's 0th column equals x_{i0}. Furthermore, the (i, j) entry of X is x_{ij}.

Renaming this tableau A, now let's bring nonbasic column 1 into the basis.

$$A_1 = 1A_2 - A_3 + 3A_4 + 10A_5 = \sum_{i=1}^{m} x_{i1} A_{B(i)}.$$

We compute

$$\theta_0 = \min_{i:x_{i1}>0} \frac{x_{i0}}{x_{i1}} = \min\left\{\frac{6}{1}, -, \frac{1}{3}, \frac{10}{10}\right\} = \frac{1}{3} = \frac{x_{30}}{x_{31}}.$$

We now want to do two things, by executing elementary row operations:

1. Update column 0, so that it contains the basic entries of the new bfs, and

2. Put e_i in column $A_{B'(i)}$, for every i. Of course, it suffices to put e_l into A_j, since the remaining e's are already where we want them.

In our example we get

	x_1	x_2	x_3	x_4	x_5	x_6
$\frac{17}{3}$	0	1	0	$-\frac{1}{3}$	0	$\frac{2}{3}$
$\frac{10}{3}$	0	0	1	$\frac{1}{3}$	0	$\frac{19}{3}$
$\frac{1}{3}$	1	0	0	$\frac{1}{3}$	0	$\frac{10}{3}$
$\frac{20}{3}$	0	0	0	$-\frac{10}{3}$	1	$-\frac{25}{3}$

The new ordered basis is $B'(1) = 2$, $B'(2) = 3$, $B'(3) = 1$, $B'(4) = 5$, column 1 having replaced column 4 in the basis.

The beauty of tableaux is that doing (2) also does (1). But this is easy to see: if $A_{B'(i)} = e_i$ for $1 \le i \le m$, then—since we haven't changed the solution space by executing elementary row operations— the ith entry of column 0 must be the $B'(i)$th entry of the bfs corresponding to B'.

Pivoting, a piece of cake. But which nonbasic column j should we bring into the basis? The cost of a bfs x_0 with ordered basis

$B(1), B(2), \ldots, B(m)$ is

$$\sum_{i=1}^{m} c_{B(i)} x_{i0}.$$

How will bringing column j into the basis change the cost? We set

$$\theta_0 = \min_{i:x_{ij}>0} \frac{x_{i0}}{x_{ij}}$$

(if the min is over a nonempty set) and choose l such that

$$\theta_0 = \frac{x_{l0}}{x_{lj}}$$

where $x_{lj} > 0$. If column j enters, $B(l)$ leaves.

$$\text{New cost} = \sum_{i\neq l} c_{B(i)}[x_{i0} - \theta_0 x_{ij}] + c_j\theta_0$$

(because $[x_{i0} - \theta_0 x_{ij}]$ will be the $B(i)$th entry of the new bfs, for $i \neq l$, and its jth entry will be θ_0).

$$\begin{aligned}
\text{Old cost} &= \sum_{i\neq l} c_{B(i)} x_{i0} + c_{B(l)} x_{l0} \\
&= \sum_{i\neq l} c_{B(i)} x_{i0} + \theta_0 c_{B(l)} x_{lj},
\end{aligned}$$

since

$$\begin{aligned}
x_{l0} &= \frac{x_{l0}}{x_{lj}} \cdot x_{lj} \\
&= \theta_0 x_{lj}.
\end{aligned}$$

So

$$\begin{aligned}
[\text{new cost} - \text{old cost}] &= \theta_0[c_j - \sum_{i\neq l} c_{B(i)} x_{ij} - c_{B(l)} x_{lj}] \\
&= \theta_0[c_j - \sum_i c_{B(i)} x_{ij}].
\end{aligned}$$

To decrease the cost, we want $c_j - \sum_i c_{B(i)} x_{ij}$ to be negative.

Where z is the n-vector satisfying $z_j = \sum_i c_{B(i)} x_{ij}$, let $\bar{c} = c - z$, the n-vector of *modified costs*. It is then profitable to bring j into the basis if $\bar{c}_j < 0$. In this case the cost increases by $\theta_0 \bar{c}_j$, which is nonpositive. (Be forewarned! If $\theta_0 = 0$ we gain nothing, but we lose nothing either.)

Trivial Observation. $(WZ)_j = W(Z_j)$ and $(WZ)^i = (W^i)Z$. In English, the jth column of a matrix product WZ is W times the jth column of Z. The ith row of WZ, as a row vector, is W^i times Z.

Lemma 14. Let X be the tableau (without column 0) after s pivots, where A was the original matrix. Let columns $B(1), B(2), \ldots, B(m)$ be the current ordered basis where column $B(i)$ of X is e_i. Let B denote the $m \times m$ matrix whose ith column is $A_{B(i)}$. Then $X = B^{-1}A$. Furthermore, the 0th column is exactly $B^{-1}b$.

Be careful. Here B represents both a sequence of integers and an $m \times m$ matrix.

Proof. X has been obtained from A by a sequence of elementary row operations.

The elementary row operation $(\text{row}_i) \leftarrow \alpha(\text{row}_i) + \beta(\text{row}_j)$ $(j \neq i)$ can be obtained by premultiplying the current matrix by

$$
\text{row } i \quad
\begin{bmatrix}
1 & 0 & \cdots & & & & \cdots & & 0 \\
0 & 1 & & & & & & & \vdots \\
\vdots & & 1 & & & & & & \\
& & & & \alpha & & \beta & & \\
& & & & & 1 & & & \\
& & & & & & 1 & & \vdots \\
\vdots & & & & & & & 1 & 0 \\
0 & \cdots & & & & \cdots & & 0 & 1
\end{bmatrix}.
$$

The α is in cell (i, i); the β, in cell (i, j). Because $\alpha \neq 0$, this matrix is invertible.

Let K_i be the matrix which accomplished the ith row operation. $X = KA$ where $K = K_s K_{s-1} \cdots K_1$. But column $B(i)$ of X is e_i, so that $(KA)_{B(i)} = e_i$.

$$(KA)_{B(i)} = K(A_{B(i)}).$$

Therefore $K(A_{B(i)}) = e_i$ for $1 \leq i \leq m$. Writing this as a matrix equation we obtain $KB = I$. Therefore $K = B^{-1}$ and $X = B^{-1}A$.

A similar argument shows that column 0 is $B^{-1}b$. ∎

Recall that z is an n-vector whose jth component

$$z_j = \sum_{i=1}^{m} x_{ij} c_{B(i)}.$$

Let c_B be the m-vector whose ith component is $c_{B(i)}$; $(c_B)_i = c_{B(i)}$. In this notation $z^T = c_B^T X = c_B^T B^{-1} A$.

So far, so good. Remember, in order to pivot we need to find a negative entry $\bar{c}_j$, but if $\bar{c} \geq 0$, then what? We're in luck: when $\bar{c} \geq 0$, the current bfs is optimal!

Theorem 15.   If $\bar{c} = c - z \geq 0$ at bfs x_0, then x_0 is optimal.

Proof.   Let y be any feasible vector, not necessarily a basic one; $Ay = b$, $y \geq 0$. Let $B(1), B(2), \ldots, B(m)$ be the current ordered basis. For all j, $c_j \geq z_j$, so $c_j y_j \geq z_j y_j$ and therefore $c^T y \geq z^T y = (c_B^T B^{-1} A)y = c_B^T B^{-1}(Ay) = c_B^T B^{-1}b$. But the cost of x_0 is

$$c^T x_0 = \sum_{i=1}^{m} c_{B(i)} x_{i0} = \sum_{i=1}^{m} c_{B(i)} (B^{-1}b)_i = c_B^T B^{-1}b \leq c^T y.$$

Hence $c^T y \geq c^T x_0$ and x_0 is optimal. ∎

Lemma 16.   For all i, $\bar{c}_{B(i)} = 0$, i.e., the modified cost of a basic column is 0.

Proof. Because $x_{t,B(i)} = 0$ if $t \neq i$, $z_{B(i)} = \sum_{t=1}^{m} x_{t,B(i)} c_{B(t)} = c_{B(i)}$ for all i. Therefore $\bar{c}_{B(i)} = 0$. ∎

Since the modified costs $\bar{c}$ are so important, why not carry them along in the computation? This is usually done in row 0 of the tableau, as follows:

1. Initially we write c_j in the $(0,j)$ position, and place a 0 in cell $(0,0)$.

Example. Min $10x_1 + 2x_2 - 4x_3 + x_4 - 2x_5$ such that

0	10	2	−4	1	−2
8	8	4	−12	0	0
45	5	2	4	11	10
21	1	1	−2	4	5

Here we've represented the original problem as a tableau.

2. Let $B(1), B(2), \ldots, B(m)$ be a basis whose basic solution is feasible. Place e_i in column $B(i)$ for each i by performing elementary row operations.

Example (cont.). Let $B(1) = 2$, $B(2) = 4$, $B(3) = 5$. After performing elementary row operations so as to put e_i in column $B(i)$, we get

0	10	2	−4	1	−2
2	2	1	−3	0	0
1	1	0	$\frac{8}{3}$	1	0
3	−1	0	$-\frac{29}{15}$	0	1

3. As yet, we still have c in row 0, not $\bar{c}$. If j is basic, $\bar{c}_j$ is 0, so we place 0's above the basic columns, by executing

$$(\text{row } 0) \leftarrow (\text{row } 0) - c_{B(i)}(\text{row } i),$$

for $i = 1, \ldots, m$. (Do this to all of row 0, including column 0.)

Example (cont.). Subtract 2·(row 1), 1·(row 2), and (-2)·(row 3) from row 0.

1	3	0	$-\frac{68}{15}$	0	0
2	2	1	-3	0	0
1	1	0	$\frac{8}{3}$	1	0
3	-1	0	$-\frac{29}{15}$	0	1

4. The jth entry of row 0 ($j \geq 1$) is now $c_j - \sum_{i=1}^{m} x_{ij} c_{B(i)} = \overline{c}_j$, the jth modified cost. In column 0 we now have $-\sum_{i=1}^{m} x_{i0} c_{B(i)} = -(\text{cost of current bfs})$.

Lemma 17. The $(0,0)$ entry will always hold the negative of the cost of the current bfs, and the $(0,j)$ entry will always hold $\overline{c}_j$ (for $j \geq 1$), if whenever we pivot we perform a row operation so as to zero out the entry atop the entering column.

Proof. Left as an exercise.

Let us review pivoting.

1. Find a column j whose row-0 entry is negative.

2. Compute $\theta_0 = \min_{i:x_{ij}>0} \dfrac{x_{i0}}{x_{ij}}$. (Assume, for now, that some $x_{ij} > 0$.)

3. Find an l such that $\theta_0 = \dfrac{x_{l0}}{x_{lj}}$ (and $x_{lj} > 0$).

Pivot on the (l,j) entry. Column j will enter the basis, $B(l)$ will leave.

What if $x_{ij} \leq 0$ for all i, thereby thwarting our attempt to define θ_0?

Theorem 18. If $\overline{c}_j < 0$ and $x_{ij} \leq 0$ for all i, then the LPS instance is unbounded.

Proof. Let $\theta \geq 0$. Let x_0 be the current bfs and let

$$w^\theta = \begin{bmatrix} w_1^\theta \\ w_2^\theta \\ \vdots \\ w_n^\theta \end{bmatrix}$$

where

$$w_k^\theta = \begin{cases} 0 & \text{if } k \neq j \text{ and } k \text{ is nonbasic,} \\ \theta & \text{if } k = j, \\ -\theta x_{ij} & \text{if } k = B(i). \end{cases}$$

Clearly, $w^\theta \geq 0$.

$$Aw^\theta = Ax_0 + \theta \left[A_j - \sum_i x_{ij} A_{B(i)} \right].$$

Where X_t is the tth column of the tableau,

$$X_j = \sum_i x_{ij} X_{B(i)}.$$

But $X_t = (KA)_t = K(A_t)$ $(t \geq 1)$ for some invertible K. Hence

$$KA_j = \sum_i x_{ij} K A_{B(i)}$$

and

$$A_j = \sum_i x_{ij} A_{B(i)}.$$

Therefore the quantity inside the square brackets above is 0, and $Aw^\theta = Ax_0 = b$ and therefore w^θ is feasible. But

$$c^T w^\theta = c^T x_0 + \theta(\underbrace{c_j - \sum_i x_{ij} c_{B(i)}}_{=\bar{c}_j < 0}).$$

So $c^T w^\theta$ can be made arbitrarily small by choosing θ sufficiently large. ∎

Example.

Minimize $19x_1 - 13x_3 + x_4 + 3x_6$

$$
\begin{aligned}
\text{s.t. } \quad 3x_1 - 4x_3 + x_4 + 4x_5 &= 6 \\
-4x_1 + x_2 + x_3 + 2x_5 &= 1 \\
5x_1 - 2x_3 - x_5 + x_6 &= 1 \\
x_i &\geq 0 \ \forall i.
\end{aligned}
$$

Equivalently,

0	19	0	−13	1	0	3
6	3	0	−4	1	4	0
1	−4	1	1	0	2	0
1	5	0	−2	0	−1	1

For the ordered basis $B(1) = 4$, $B(2) = 2$, and $B(3) = 6$, we already have e_i in column $B(i)$ and a 0 above column 2, but not above columns 4 or 6. We subtract 1·(row 1) and then 3·(row 3) from row 0 in order to zero out the fourth and sixth entries in row 0.

−9	1	0	−3	0	−1	0
6	3	0	−4	1	4	0
1	−4	1	1	0	2	0
1	5	0	−2	0	−1	1

Now we choose, say, $j = 5$, and then $l = 2$. We pivot on the 2 and zero out the fifth entry of row 0.

$-\frac{17}{2}$	−1	$\frac{1}{2}$	$-\frac{5}{2}$	0	0	0
4	11	−2	−6	1	0	0
$\frac{1}{2}$	−2	$\frac{1}{2}$	$\frac{1}{2}$	0	1	0
$\frac{3}{2}$	3	$\frac{1}{2}$	$-\frac{3}{2}$	0	0	1

Choosing, say, $j = 3$ and then $l = 2$, we now pivot on the $(2,3)$ entry.

-6	-11	3	0	0	5	0
10	-13	4	0	1	12	0
1	-4	1	1	0	2	0
3	-3	2	0	0	3	1

Choose $j = 1$. We conclude from the negativity of $\overline{c}_j$ and the nonpositive numbers in column j that the objective function can be made

arbitrarily small. Indeed, $\begin{bmatrix} \theta \\ 0 \\ 1+4\theta \\ 10+13\theta \\ 0 \\ 3+3\theta \end{bmatrix}$ is feasible for all $\theta \geq 0$ and

of cost $6 - 11\theta$.

Note that pivoting takes only $O(mn)$ time.

2.3   Cycling

How could the Simplex Algorithm fail to terminate? We start with a vertex v_0, and build a string of vertices $v_1, v_2, v_3, \ldots$ of nonincreasing cost. If we discover that v_i is optimal, or that the problem is unbounded, we halt. Since we obviously can't construct an infinite sequence of vertices of strictly decreasing cost, the only way Simplex can fail to halt is for it to get stuck at a vertex v_i indefinitely, running *ad infinitum* through a string of bases all of which correspond to v_i. Such *cycling* can indeed occur (though extremely rarely in practice).

Since the basis determines the tableau, provided that we use some fixed tie-breaking rule, if we ever repeat a basis we will loop forever. To prove Simplex terminates it is enough somehow to ensure that we never repeat a basis.

Bland's Rule, a way of selecting the entering and exiting columns, is this: Choose the smallest numbered column as possible to enter the basis, and then, the smallest possible numbered column to leave it. More precisely, always choose j minimal such that $\bar{c}_j < 0$, and afterward, choose an l for which $B(l)$ is as small as possible, from among all l satisfying $x_{lj} > 0$ and $\dfrac{x_{l0}}{x_{lj}} = \theta_0$. (Be careful. Choose l so that $B(l)$ is minimum, not l itself.) Bland's Rule guarantees termination:

Theorem 19. Simplex terminates as long as the pivot row and column are chosen by Bland's Rule.

Proof. Suppose we have a linear program that cycles when we use Bland's Rule for pivoting. We will show, first, that there is a simpler linear program that also cycles, and second, that the simpler linear program cannot cycle.

Suppose Bland's Rule gives rise to a cycle. When we pivot, we bring a new column into the basis in exchange for one of the old ones; say a column is a *pivot column* and a row a *pivot row* if, at some point in the cycle, we pivot on an entry in that row or column. Let column q be the highest-numbered pivot column. Suppose the cycle takes us through tableaux $T = T_1, T_2, ..., T_r = T_1$, where T is chosen so that in passing from $T = T_1$ to T_2, column q enters the basis (q is the "entering column"). At some later time in the cycle, column q leaves the basis; let T' be a tableau within which column q is basic, but for which the next tableau contains column $p < q$ in place of q (q is "exiting").

If a nonpivot column is deleted from all the tableaux, the algorithm will pivot just as before. Delete all nonpivot columns (other than column 0) from all the tableaux. Furthermore, if we delete a nonpivot row, again we will pivot exactly as before. The elementary row operations make this less clear, but we note that within the cycle, we subtract multiples of pivot rows from nonpivot rows, but we never add multiples of nonpivot rows to pivot rows. The pivot rows don't "depend on" nonpivot rows. Delete all nonpivot rows from all the tableaux in the cycle. As all the pivots in the cycle are degenerate, all entries of column 0 (other than the 0th) are now 0.

Thanks to the 0's below the $(0,0)$ cell, whatever number occupies the $(0,0)$ entry of T occupies it in all tableaux. Only a bookkeeping aid, the $(0,0)$ entry does not affect how we pivot. Our last change is to change that entry in all tableaux to 0.

Now, focus on T. Let $T = (x_{ij})$, $0 \leq i \leq m$, $0 \leq j \leq n$, and let c be the n-vector whose transpose appears in row 0. Let the ordered basis be $B(1), B(2), ..., B(m)$. Imagine that T represents the initial tableau for a linear program with 0 right-hand side and objective function $c^T x$.

Let $T' = (x'_{ij})$, $0 \leq i \leq m$, $0 \leq j \leq n$, let c' be the n-vector whose transpose appears in row 0, and let $B'(1), B'(2), ..., B'(m)$ be its ordered basis, which contains q but not p. Let y be the n-vector defined as follows:

- $y_p = 1$.
- $y_{B'(i)} = -x'_{ip}$, for $i = 1, 2, ..., m$.
- All other entries of y are 0.

Vector y is 0 in all nonbasic positions except the pth.

On the ith row of T' $(1 \leq i \leq m)$, the only nonzeroes are in the nonbasic columns, except that $x'_{i,B'(i)} = 1$. In all nonbasic columns j other than the pth, y_j is 0. So the dot product of y and the ith row of T' is precisely

$$x'_{ip} \cdot y_p + x'_{i,B'(i)} \cdot y_{B'(i)} = x'_{ip} \cdot 1 + 1 \cdot (-x'_{ip}) = 0.$$

Thus y satisfies the constraint implied by the ith row.

On the 0th row of T', we have zeroes above all the basic columns, so

$$c'^T y = x'_{0p} \cdot y_p = x'_{0p} < 0;$$

x'_{0p} is negative because column p is entering the basis. As we pivot through the cycle, we subtract from the 0th row multiples of the other rows. However, since y satisfies all of the constraints implied by the lower rows of T', not only is the dot product of y and the 0th row of T' equal to x'_{0p}, but so is the dot product of y and the 0th row of any of the tableaux T_k! Thus,

$$c^T y = c'^T y < 0.$$

What can we say about c, the transpose of the 0th row of T? Because column q is entering in T, $c_q < 0$, and, since we chose the lowest-numbered column in T to enter the basis, $c_j \geq 0$ if $j < q$.

Note that $y_q < 0$, because column q is leaving the basis, and hence $x'_{ip} > 0$. If column j is nonbasic in T', $y_j \geq 0$. Now suppose that there is an l, $1 \leq l \leq m$, such that $B'(l) \neq q$ yet $y_{B'(l)} = -x'_{lp} < 0$. From $y_{B'(l)} < 0$ we infer that $x'_{lp} > 0$. Now, since the 0th column is all 0, we could have pivoted on any positive entry of column p. Since $B'(l) < q$, according to the second half of Bland's Rule we should have evicted column $B'(l)$, not column q, a contradiction. Thus $y_j \geq 0$ for all $j < q$.

It follows that

$$c^T y = c_q y_q + \sum_{j<q} c_j y_j \geq c_q y_q > 0,$$

contradicting the fact that $c^T y < 0$. ∎

2.4 Getting Started

The only remaining gap is our description of Simplex is the question of getting started. If we can find a feasible point—if there *are* any at all—we can find a bfs of no greater cost.

If we start with $\min c^T x$ and if $b \geq 0$, we're in luck. Convert the
$$Ax \leq b$$
$$x \geq 0$$
problem to standard form by adding n nonnegative "slack variables." They form a bfs, and the corresponding columns are a basis.

Example.

$$\begin{array}{c} \min c^T x \\ x_1 + 2x_2 \le 5 \\ 3x_1 - x_2 \le 22 \\ x_1, x_2 \ge 0 \end{array} \quad \longrightarrow \quad \begin{array}{c} \min c^T x \\ x_1 + 2x_2 + x_3 + 0x_4 = 5 \\ 3x_1 - x_2 + 0x_3 + x_4 = 22 \\ x_i \ge 0 \; \forall i \end{array}$$

has basis $\{3,4\}$ and bfs $\begin{bmatrix} 0 \\ 0 \\ 5 \\ 22 \end{bmatrix}$.

When no bfs is evident, we can use the "artificial-variable" or "two-phase" method. Let $\min c^T x$ be our initial problem. By

$$Ax = b$$
$$x \ge 0$$

negating some constraints, we may assume without loss of generality that $b \ge 0$. Here is the two-phase method.

First reduce $\boxed{A \mid b}$ to row-echelon form. If the rank of A is r yet the ith row of A for some $i > r$ has a nonzero right-hand side (and all-zero left-hand side), report infeasibility and halt. If no such rows exist, eliminate the all-zero rows. Call this new matrix A and let b be the new right-hand side. If A is $m \times n$, $\text{rank}(A) = m$.

(Phase I) Add m new "artificial" variables x_i', $1 \le i \le m$:

$$\begin{array}{rcl} x_1' \qquad\qquad\qquad + a_{11}x_1 + a_{12}x_2 + \cdots + a_{1n}x_n &=& b_1 \\ x_2' \qquad\qquad + a_{21}x_1 + a_{22}x_2 + \cdots + a_{2n}x_n &=& b_2 \\ \ddots \qquad\qquad\qquad\qquad\qquad\qquad &\vdots& \\ x_m' + a_{m1}x_1 + a_{m2}x_2 + \cdots + a_{mn}x_n &=& b_m \end{array}$$

$$x_i' \ge 0 \; \forall i, \; x_j \ge 0 \; \forall j.$$

Minimize $\sum_{i=1}^{m} x_i'$, starting with the bfs in which $x_i' = b_i$ for $i = 1, 2, ..., m$.

Lemma 20. This new problem has optimum value 0 if and only if the original problem is feasible.

Proof. Clear. ∎

If the optimal value found in Phase I is strictly positive, report infeasibility and halt. Otherwise, where $x_1^*, x_2^*, \ldots, x_n^*$ is the sequence of values of *original* variables returned by the algorithm, $x^* = [x_1^* \; x_2^* \; \cdots \; x_n^*]^T$ is a feasible point in our original problem.

If the Phase I linear program was solved by the Simplex Algorithm, then x^* is a vertex of our original problem (why?); find a basis of A corresponding to x^*. If the Phase I linear program was not solved via Simplex, find a bfs v of no greater cost than that of x^* and a basis corresponding to v.

(Phase II) Solve

$$\min c^T x$$
$$\text{s.t. } Ax = b$$
$$x \geq 0$$

starting with the basis obtained in Phase I. (A is $m \times n$ and of rank m.)

2.5 Worst-Case Performance

How many pivots does Simplex execute? Most pivoting rules lead to variants of Simplex that require exponential time in the worst case. For example, let n be a positive integer and consider the linear program

$$\max \sum_{j=1}^{n} 10^{n-j} x_j$$
$$\text{s.t. } 2\sum_{j=1}^{i-1} 10^{i-j} x_j + x_i \leq 100^{i-1}, \text{ for } 1 \leq i \leq n$$
$$x_j \geq 0 \; \forall j.$$

The $n = 3$ case is

$$\max 100x_1 + 10x_2 + x_3$$

$$
\begin{array}{rrcl}
\text{s.t.} & x_1 & \leq & 1 \\
& 20x_1 + x_2 & \leq & 100 \\
& 200x_1 + 20x_2 + x_3 & \leq & 10000 \\
& x_1, x_2, x_3 & \geq & 0
\end{array}
$$

The initial bfs is $[0 \ \ 0 \ \ 0 \ \ 1 \ \ 100 \ \ 10000]^T$.

If we pivot by choosing the entering variable j such that $\bar{c}_j$ is most negative, Simplex will go through $2^n - 1$ pivots!

There are examples known for almost all known pivoting strategies that force Simplex to execute an exponential number of pivots. Suggesting a new pivoting strategy is a lot easier than proving that it requires exponential time.

It is an open question whether there is *any* pivoting rule that guarantees termination after a polynomial number of pivots. This related question is also open: does there exist a polynomial $p(m, n)$ such that whenever $\mathcal{B}$ and $\mathcal{B}'$ are two bases whose basic solutions are feasible, there is a sequence of at most $p(m, n)$ pivots that will convert the bfs for $\mathcal{B}$ to the bfs for $\mathcal{B}'$, *while maintaining feasibility throughout?* (If we drop the "maintaining feasibility" condition, the answer is yes. It suffices to take $p(m, n) = m$.) If the answer is no, no variant of Simplex could ever run in polynomial time.

How well does Simplex perform on *real* problems? Remarkably well. In practice, the number of pivots seems to be between $4m$ and $6m$ for Phase I and Phase II together. Rarely does either phase require more than $10m$ pivots. As n grows, the number of pivots seems to grow slowly, perhaps logarithmically, with n.

■ Simplex Algorithm

2.6 Notes

Between the creation of the Simplex Algorithm by G. Dantzig in 1947 [17] and the appearance of the Ellipsoid Algorithm in 1979 [35], and to a lesser extent afterward, paper after paper appeared on the Simplex Algorithm. Few algorithms have been studied more extensively. Several researchers showed that Simplex will run quickly on average, provided that the inputs are drawn from a given probability distribution. See the survey by Shamir [48], the book by Borgwardt [13], the paper by Adler [1], those by Adler, Karp and Shamir [2, 3], the one by Adler and Megiddo [4], and those by Smale [54, 55]. The claims about the performance of Simplex on real-world problems are gleaned from Shamir's survey [48].

The notation of this chapter is from Papadimitriou and Steiglitz [44]; likewise for the statements of Theorem 13 and Lemmas 16 and 17. Our proof of Bland's Theorem is based on that in Kuhn's lecture notes [37], which differs from Bland's original proof [11].

The exponential-time instances of Simplex are due to Klee and Minty [36], as presented by Chvátal [16]. Klee and Minty were the first to prove that a variant of Simplex could take exponential time, and Chvátal [8] proved that Simplex requires exponentially many pivots in the worst case, if Bland's pivoting rule is used. Jeroslow [32] proved the same for the "largest possible improvement" pivot rule. The first proof that cycling could actually occur is due to Hoffman [30].

The reader can find the Simplex Algorithm in virtually every book that covers linear programming. Among them are those by Calvert and Voxman [15], Chvátal [16], Gass [22], Glicksman [23], Hadley [28], Papadimitriou and Steiglitz [44], Simonnard [53] and Strang [56].

Chapter 3

Duality

3.1 The Dual

Let us take a linear program in standard form and try to derive *lower bounds* on the optimal cost (if it exists).

$$\begin{array}{rl}
\text{Min} & x_1 + 2x_2 - 7x_3 \\
\text{s.t.} & 3x_1 - x_2 + 5x_3 \;=\; 6 \\
& 2x_1 + 0x_2 - x_3 \;=\; 12
\end{array}$$

$$x_1 \geq 0 \quad x_2 \geq 0 \quad x_3 \geq 0$$

Let $w = \begin{bmatrix} w_1 \\ w_2 \\ w_3 \end{bmatrix}$ be any feasible point and let $y = \begin{bmatrix} y_1 \\ y_2 \end{bmatrix}$ be arbitrary. Consider linear combinations of the constraints, using the y's as coefficients, with x_i replaced by w_i.

Let $f = y_1 \underbrace{(3w_1 - w_2 + 5w_3)}_{=6} + y_2 \underbrace{(2w_1 - w_3)}_{=12} = 6y_1 + 12y_2.$

$$f = (3y_1 + 2y_2)w_1 + (-y_1 + 0y_2)w_2 + (5y_1 - y_2)w_3$$

49

If

$$3y_1 + 2y_2 \leq c_1 = 1$$
$$-y_1 + 0y_2 \leq c_2 = 2 \tag{3.1}$$
$$5y_1 - y_2 \leq c_3 = -7,$$

then, because $w \geq 0$, we may conclude that

$$
\begin{aligned}
f &= 6y_1 + 12y_2 \\
&\leq 1 \cdot w_1 + 2 \cdot w_2 - 7 \cdot w_3 \quad \text{(because } w \geq 0\text{)} \\
&= c^T w.
\end{aligned}
$$

To recap, if $\left\{ \begin{aligned} 3y_1 + 2y_2 &\leq & 1 \\ -y_1 + 0y_2 &\leq & 2 \\ 5y_1 - y_2 &\leq & -7 \end{aligned} \right.$, then $6y_1 + 12y_2 \leq c^T w$. In particular, if w^* is optimal in the original problem and (3.1) holds, then

$$6y_1 + 12y_2 \leq c^T w^*.$$

For example, $\begin{bmatrix} -2 \\ 0 \end{bmatrix}$ satisfies (3.1). Therefore $-12 \leq c^T w^*$.

Finding the best lower bound in this way is itself naturally phrased as a linear program:

$$
\begin{aligned}
\max \quad & 6y_1 + 12y_2 \\
\text{s.t.} \quad & 3y_1 + 2y_2 \leq 1 \\
& -y_1 + 0y_2 \leq 2 \\
& 5y_1 - y_2 \leq -7 \\
& y_1 \gtrless 0 \quad y_2 \gtrless 0,
\end{aligned}
$$

which can be equivalently written in this bizarre way:

$1 \geq$	$2 \geq$	$-7 \geq$	max	
$3y_1$	$-y_1$	$5y_1$	$6y_1$	$y_1 \gtrless 0$
$+2y_2$	$+0y_2$	$-y_2$	$+12y_2$	$y_2 \gtrless 0$

By comparison, the original problem is

$$
\begin{array}{llll}
\min & 1x_1 & +2x_2 & -7x_3 \\
& 3x_1 & -1x_2 & +5x_3 & = 6 \\
& 2x_1 & +0x_2 & -1x_3 & = 12
\end{array}
$$

$$ x_1 \geq 0 \quad x_2 \geq 0 \quad x_3 \geq 0. $$

The numbers match!

Let's try doing the same thing for a linear program in general form:

$$
\begin{array}{l}
\min \quad 3x_1 - x_2 - 4x_3 \\
\quad 5x_1 - 6x_2 + 7x_3 = 9 \\
\quad 6x_1 - 9x_2 - 2x_3 \geq -7
\end{array}
$$

$$ x_1 \geq 0 \quad x_2 \gtrless 0 \quad x_3 \gtrless 0 $$

Let w be a feasible point ($w_1 \geq 0$, but w_2 and w_3 can be positive or negative). Let $y_1, y_2 \in \mathbb{R}$.

Consider $f = y_1 \underbrace{(5w_1 - 6w_2 + 7w_3)}_{=9} + y_2 \underbrace{(6w_1 - 9w_2 - 2w_3)}_{\geq -7}$.

$$ f = (5y_1 + 6y_2)w_1 + (-6y_1 - 9y_2)w_2 + (7y_1 - 2y_2)w_3 $$

If $y_2 \geq 0$, then $f \geq 9y_1 - 7y_2$. If $(5y_1 + 6y_2) \leq 3$, $(-6y_1 - 9y_2) = -1$, $(7y_1 - 2y_2) = -4$ and $y_2 \geq 0$, then

$$
\begin{array}{rl}
c^T w & = \quad 3w_1 - w_2 - 4w_3 \\
& \geq \quad (5y_1 + 6y_2)w_1 + (-6y_1 - 9y_2)w_2 + (7y_1 - 2y_2)w_3 \\
& = \quad f \\
& \geq \quad 9y_1 - 7y_2.
\end{array}
$$

Note that because x_1 is sign-constrained, we require only that $5y_1 + 6y_2 \leq 3$, whereas because x_2 and x_3 are not sign-constrained, we must require $-6y_1 - 9y_2 = -1$ and $7y_1 - 2y_2 = -4$ (equalities instead of inequalities). Similarly, because the second constraint is an equality, y_1 is not sign-constrained, unlike the sign-constrained y_2. This is important!

To recap, $9y_1 - 7y_2 \leq c^T w$ if

$$
\begin{aligned}
5y_1 + 6y_2 &\leq 3 \\
-6y_1 - 9y_2 &= -1 \\
7y_1 - 2y_2 &= -4 \\
y_1 \gtrless 0 \quad y_2 &\geq 0.
\end{aligned}
$$

To get the best lower bound on the objective function of the original problem, we should solve

$$
\begin{aligned}
\max \quad & 9y_1 - 7y_2 \\
\text{s.t.} \quad & 5y_1 + 6y_2 \leq 3 \\
& -6y_1 - 9y_2 = -1 \\
& 7y_1 - 2y_2 = -4 \\
& y_1 \gtrless 0 \quad y_2 \geq 0.
\end{aligned}
$$

Once again, we see that the coefficients of the jth constraint come from column j of the original problem. The right-hand sides come from the original cost coefficients, and the cost function comes from the original right-hand side. This new problem is known as the *DUAL* of the original problem; the original problem is known as the *PRIMAL*.

Here is how we build the dual in general. For each *constraint* in the primal, there is a *variable* in the dual. For each *variable* in the primal, there is a *constraint* in the dual.

	PRIMAL	DUAL	
row i	$\sum_j a_{ij}x_j = b_i$	$y_i \gtrless$	0
row i	$\sum_j a_{ij}x_j \geq b_i$	$y_i \geq$	0
var j	$x_j \gtrless 0$	$\sum_{i=1}^{m} y_i a_{ij} =$	c_j
var j	$x_j \geq 0$	$\sum_i y_i a_{ij} \leq$	c_j
min	$c^T x$	max	$y^T b$

To get the dual, a maximization problem, first turn any $\leq$ constraints in the primal into $\geq$'s by negating both sides. Put the coefficients of the objective function on the right-hand side of the dual. Read down the primal's columns and use the entries in one column to write down one dual constraint. A dual variable y_i is sign-constrained ($y_i \geq 0$) if and only if the corresponding primal constraint is an inequality; a dual constraint is an inequality if and only if the corresponding primal variable is sign-constrained ($x_j \geq 0$). Remember, inequality in one problem corresponds to inequality in the other. Equality in one corresponds to $\gtrless$ in the other.

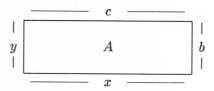

Example.

$$
\begin{array}{rl}
\text{PRIMAL:} \quad \min & 3x_1 - 6x_2 - x_3 + 2x_4 \\
& 2x_1 - 16x_2 + 12x_3 + 0x_4 \;\geq\; 12 \\
& 5x_1 + 5x_2 + 0x_3 - 13x_4 \;\leq\; 23 \\
& -4x_1 - 3x_2 - x_3 - 7x_4 \;=\; 28 \\[4pt]
& x_1 \geq 0 \quad x_2 \gtrless 0 \quad x_3 \gtrless 0 \quad x_4 \geq 0
\end{array}
$$

In matrix form:

min	3	-6	-1	2		
	2	-16	12	0	$\geq$	12
	-5	-5	0	13	$\geq$	-23
	-4	-3	-1	-7	$=$	28
	$\geq$	$\gtrless$	$\gtrless$	$\geq$		

$$
\begin{array}{rl}
\text{DUAL:} \quad \max & 12y_1 - 23y_2 + 28y_3 \\
& 2y_1 - 5y_2 - 4y_3 \;\leq\; 3 \\
& -16y_1 - 5y_2 - 3y_3 \;=\; -6 \\
& 12y_1 + 0y_2 - y_3 \;=\; -1 \\
& 0y_1 + 13y_2 - 7y_3 \;\leq\; 2 \\[4pt]
& y_1 \geq 0 \quad y_2 \geq 0 \quad y_3 \gtrless 0
\end{array}
$$

In matrix form:

max	12	−23	28		
	2	−5	−4	≤	3
	−16	−5	−3	=	−6
	12	0	−1	=	−1
	0	13	−7	≤	2
	≥	≥	$\gtreqless$		

Theorem 21. The dual of the dual is the primal.

Proof. We can renumber the constraints and relabel the variables so that the primal becomes

$$
\begin{aligned}
\min \; & c^T x \\
A^i x \; &= \; b_i, \quad 1 \le i \le h \\
A^i x \; &\ge \; b_i, \quad h+1 \le i \le m \\
x_j \; &\ge \; 0, \quad 1 \le j \le l \\
x_j \; &\gtreqless \; 0, \quad l+1 \le j \le n.
\end{aligned}
$$

The dual is

$$
D: \quad
\begin{aligned}
\max \; & y^T b \\
y_i \; &\gtreqless \; 0, \quad 1 \le i \le h \\
y_i \; &\ge \; 0, \quad h+1 \le i \le m \\
(A^T)^j y = A_j^T y \; &\le \; c_j, \quad 1 \le j \le l \\
(A^T)^j y = A_j^T y \; &= \; c_j, \quad l+1 \le j \le n.
\end{aligned}
$$

Write the dual as a minimization problem:

$$
D: \quad
\begin{aligned}
\min \; & y^T(-b) \\
y_i \; &\gtreqless \; 0, \quad 1 \le i \le h \\
y_i \; &\ge \; 0, \quad h+1 \le i \le m \\
(-A^T)^j y \; &\ge \; -c_j, \quad 1 \le j \le l \\
(-A^T)^j y \; &= \; -c_j, \quad l+1 \le j \le n.
\end{aligned}
$$

The dual of D is

$$
\begin{array}{rcl}
\max x^T(-c) & & \\
((-A^T)^T)^i x & = & -b_i, \quad 1 \leq i \leq h \\
((-A^T)^T)^i x & \leq & -b_i, \quad h+1 \leq i \leq m \\
x_j & \geq & 0, \quad 1 \leq j \leq l \\
x_j & \gtrless & 0, \quad l+1 \leq j \leq n,
\end{array}
$$

which is the original problem. ∎

3.2 The Duality Theorem

From our construction of the dual, a reasonable conjecture would be that if w is feasible in the primal and u is feasible in the dual, then $c^T w \geq u^T b$. It *is* true. (If A is $m \times n$, c and w are n-vectors and u and b are m-vectors.)

Theorem 22. If w, u is a primal/dual feasible pair, $c^T w \geq u^T b$.

Proof. We use the general-form linear program in the form of Theorem 21.

$$
c^T w = \sum_{j=1}^{l} c_j \cdot w_j + \sum_{j=l+1}^{n} c_j \cdot w_j,
$$

where $c_j \geq A_j^T u$ and $w_j \geq 0$ for $j \leq l$, and $c_j = A_j^T u$ and $w_j \gtrless 0$ for $j > l$. Where $A = (a_{ij})$,

$$
\begin{aligned}
c^T w & \geq \sum_{j=1}^{l} (A_j^T u) w_j + \sum_{j=l+1}^{n} (A_j^T u) w_j \\
& = \sum_{i=1}^{m} u_i \left[\sum_{j=1}^{l} a_{ij} w_j + \sum_{j=l+1}^{n} a_{ij} w_j \right].
\end{aligned}
$$

The quantity in brackets is $A^i w$. Thus

$$
c^T w \geq \sum_{i=1}^{h} u_i [A^i w] + \sum_{i=h+1}^{m} u_i [A^i w].
$$

Since $A^i w = b_i$ in the first summation, and since in the second $u_i \geq 0$ and $A^i w \geq b_i$, $c^T w \geq \sum_{i=1}^m u_i b_i = u^T b$. ∎

We now prove the extremely important **Duality Theorem**.

Theorem 23. If a linear program has an optimal solution, so does its dual, and their optimal costs are identical.

A lot of mathematics (e.g., combinatorial optimization, mathematical economics) is based on this theorem!

Proof Sketch. We prove the statement only for linear programs in the LPS form. Let the primal P and the dual D be respectively
min $c^T x$ and   max $y^T b$.
$Ax = b$ $A^T y \leq c$
$x \geq 0$ $y \gtrless 0$

If P has an optimal point, Simplex will find it. Let v^* be the primal optimal bfs discovered by Simplex, and let $B(1), B(2), \ldots, B(m)$ be an ordered basis corresponding to v^*. Let B be the $m \times m$ matrix whose ith column is $A_{B(i)}$, and let c_B be the m-vector whose ith component is $c_{B(i)}$. At termination, the 0th row $\bar{c}^T$ is nonnegative. Recall that $\bar{c} = c - z$ where $z^T = c_B^T B^{-1} A$. Hence at termination $c - A^T (B^{-1})^T c_B \geq 0$.

Let $u^* = (B^{-1})^T c_B$, an m-vector. Then $c - A^T u^* \geq 0$ and therefore $A^T u^* \leq c$; consequently u^* is dual feasible. But

$$c^T v^* = c_B^T \cdot \underbrace{(\text{column 0 of final tableau})}_{=B^{-1}b}.$$

So

$$c^T v^* = c_B^T B^{-1} b = (u^*)^T b,$$

where $c^T v^* = $ [cost of v^* in the primal], and $(u^*)^T b = $ [cost of u^* in the dual].

Since for any dual feasible u, $c^T v^* \geq u^T b$, this u^* must be dual optimal. ∎

Corollary 24. If we run Simplex on LPS, at termination $(B^{-1})^T c_B$ is dual optimal (if an optimal point exists).

Example. Consider this linear program:

$$P:\quad
\begin{array}{|c|ccccc|}
\hline
0 & 0 & 0 & -1 & 0 & -3 \\
\hline
13 & 1 & 0 & \frac{5}{2} & 0 & 1 \\
32 & 1 & 0 & 7 & 1 & 5 \\
28 & 2 & 1 & \frac{11}{2} & 0 & 4 \\
\hline
\end{array}
\quad \leftarrow c^T$$

$$x_j \geq 0 \quad \forall j.$$

A basis is $B(1) = 1$, $B(2) = 4$, $B(3) = 2$ and this basis yields a bfs:

$$
\begin{array}{|c|ccccc|}
\hline
0 & 0 & 0 & -1 & 0 & -3 \\
\hline
13 & 1 & 0 & \frac{5}{2} & 0 & 1 \\
19 & 0 & 0 & \frac{9}{2} & 1 & 4 \\
2 & 0 & 1 & \frac{1}{2} & 0 & 2 \\
\hline
\end{array}
$$

Fortunately, $\bar{c} = c$ is already in row 0, so we can choose $j = 3$ and pivot on the $\frac{1}{2}$:

$$
\begin{array}{|c|ccccc|}
\hline
4 & 0 & 2 & 0 & 0 & 1 \\
\hline
3 & 1 & -5 & 0 & 0 & -9 \\
1 & 0 & -9 & 0 & 1 & -14 \\
4 & 0 & 2 & 1 & 0 & 4 \\
\hline
\end{array}
$$

which is an optimal tableau. An optimal ordered basis is given by

$$B(1) = 1, \quad B(2) = 4, \quad B(3) = 3,$$ and the corresponding bfs,
$$\begin{bmatrix} 3 \\ 0 \\ 4 \\ 1 \\ 0 \end{bmatrix},$$

of cost -4, is optimal.

The dual D of P is

$$\begin{aligned}
\max \quad & 13y_1 + 32y_2 + 28y_3 \\
\text{s.t.} \quad & y_1 + y_2 + 2y_3 & \leq & \quad 0 \\
& y_3 & \leq & \quad 0 \\
& \tfrac{5}{2}y_1 + 7y_2 + \tfrac{11}{2}y_3 & \leq & \quad -1 \\
& y_2 & \leq & \quad 0 \\
& y_1 + 5y_2 + 4y_3 & \leq & \quad -3 \\
& y_i & \gtrless & \quad 0 \quad \forall i.
\end{aligned}$$

Since $B = \begin{bmatrix} 1 & 0 & \frac{5}{2} \\ 1 & 1 & 7 \\ 2 & 0 & \frac{11}{2} \end{bmatrix}$ and $c_B = \begin{bmatrix} 0 \\ 0 \\ -1 \end{bmatrix}$, a dual optimal vector is

$$\left\{ \begin{bmatrix} 1 & 0 & \frac{5}{2} \\ 1 & 1 & 7 \\ 2 & 0 & \frac{11}{2} \end{bmatrix}^{-1} \right\}^T \begin{bmatrix} 0 \\ 0 \\ -1 \end{bmatrix} = \begin{bmatrix} 11 & 17 & -4 \\ 0 & 1 & 0 \\ -5 & -9 & 2 \end{bmatrix} \begin{bmatrix} 0 \\ 0 \\ -1 \end{bmatrix} = \begin{bmatrix} 4 \\ 0 \\ -2 \end{bmatrix}.$$

Indeed, $\begin{bmatrix} 4 \\ 0 \\ -2 \end{bmatrix}$ is dual feasible, and, having dual cost -4, is also dual optimal.

If the primal is unbounded, the dual must infeasible. Otherwise, how could $c^T w \geq u^T b$, if $c^T w$ can be made arbitrarily small? Analogously, if the dual is unbounded, the primal is infeasible.

Corollary 25. Exactly one of these three cases occurs:

1. Primal and dual are both infeasible.

2. One is unbounded and the other is infeasible.

3. Both have optimal points.

Proof. Clear.

In fact, each of these three can occur.

Corollary 26. Suppose that the leftmost m columns of A are an identity matrix. If we run Simplex on LPS and it terminates at an optimal point, at termination the m-vector u^* given by $u_i^* = c_i - \bar{c}_i$ is dual optimal.

Proof. At optimality $\bar{c} = c - A^T(B^{-1})^T c_B$. For $1 \leq i \leq m$,

$$(c - \bar{c})_i = [A^T(B^{-1})^T c_B]_i \ = \ A_i^T(B^{-1})^T c_B$$
$$= \ [(B^{-1})^T c_B]_i,$$

because $A_i = e_i$. So u^* given by $u_i^* = (c - \bar{c})_i$ is optimal by Corollary 24. ∎

Example. Let $P =$

0	1	1	1	4	−1
33	1	0	0	1	−3
220	0	1	0	4	−1
253	0	0	1	1	2

$$x_j \geq 0 \quad \forall j.$$

Take $B(1) = 1$, $B(2) = 2$, $B(3) = 3$, and place 0's above columns 1, 2, and 3:

−506	0	0	0	−2	1
33	1	0	0	1	−3
220	0	1	0	4	−1
253	0	0	1	1	2

Choose $j = 4$ and pivot on the $(1,4)$ entry:

−440	2	0	0	0	−5
33	1	0	0	1	−3
88	−4	1	0	0	11
220	−1	0	1	0	5

Choose $j = 5$ and pivot on the 11:

-400	$\frac{2}{11}$	$\frac{5}{11}$	0	0	0
57	$-\frac{1}{11}$	$\frac{3}{11}$	0	1	0
8	$-\frac{4}{11}$	$\frac{1}{11}$	0	0	1
180	$\frac{9}{11}$	$-\frac{5}{11}$	1	0	0

This tableau is optimal. The cost of $\begin{bmatrix} 0 \\ 0 \\ 180 \\ 57 \\ 8 \end{bmatrix}$ is 400.

The dual D of P is

$$
\begin{array}{rl}
\max & 33y_1 + 220y_2 + 253y_3 \\
\text{s.t.} & \\
& y_1 \leq 1 \\
& y_2 \leq 1 \\
& y_3 \leq 1 \\
& y_1 + 4y_2 + y_3 \leq 4 \\
& -3y_1 - y_2 + 2y_3 \leq -1 \\
& y_1 \gtrless 0 \quad y_2 \gtrless 0 \quad y_3 \gtrless 0.
\end{array}
$$

The leftmost three columns of A are an identity matrix. By Corollary 25, $u^* = \begin{bmatrix} 1 - \frac{2}{11} \\ 1 - \frac{5}{11} \\ 1 - 0 \end{bmatrix} = \begin{bmatrix} \frac{9}{11} \\ \frac{6}{11} \\ 1 \end{bmatrix}$ should be dual optimal. Indeed, u^* is feasible in D, and since its dual cost is 400, it is dual optimal as well.

3.3   Complementary Slackness

We now prove a very important theorem that allows us to determine if a pair of vectors, respectively primal and dual feasible, are primal and dual optimal.

Theorem 27. (Complementary Slackness)

Let P be a linear program in general form:

$$
\begin{array}{rll}
P: & \min c^T x & \\
& \text{s.t. } A^i x = b_i, & 1 \le i \le h \\
& A^i x \ge b_i, & h+1 \le i \le m \\
& x_j \ge 0, & 1 \le j \le l \\
& x_j \gtrless 0, & l+1 \le j \le n.
\end{array}
$$

Let its dual be D :

$$
\begin{array}{rll}
& \max y^T b & \\
& \text{s.t. } y_i \gtrless 0, & 1 \le i \le h \\
& y_i \ge 0, & h+1 \le i \le m \\
& A_j^T y \le c_j, & 1 \le j \le l \\
& A_j^T y = c_j, & l+1 \le j \le n.
\end{array}
$$

Let w be primal feasible and let u be dual feasible. Then w is primal optimal and u is dual optimal if and only if

$$(A^i w - b_i) u_i = 0 \text{ for } i = 1, 2, ..., m$$

and

$$w_j (c_j - A_j^T u) = 0 \text{ for } j = 1, 2, ..., n.$$

(If a dual variable is nonzero, the corresponding primal constraint must be tight. If a primal variable is nonzero, the corresponding dual constraint must be tight.)

Proof. Note that $(A^i w - b_i) u_i \ge 0$ for all i and $w_j(c_j - A_j^T u) \ge 0$ for all j. Let

$$t = \sum_{i=1}^{m} (A^i w - b_i) u_i + \sum_{j=1}^{n} w_j(c_j - A_j^T u).$$

Then $t = 0$ if and only if

$$(A^i w - b_i) u_i = 0 \text{ for all } i$$

and

$$w_j(c_j - A_j^T u) = 0 \text{ for all } j.$$

$$t = -\sum_{i=1}^{m} b_i u_i + \sum_{j=1}^{n} w_j c_j + \sum_{i=1}^{m} u_i(A^i w) - \sum_{j=1}^{n} w_j(A_j^T u)$$

But

$$\sum_{i=1}^{m} u_i(A^i w) - \sum_{j=1}^{n} w_j(A_j^T u) = \sum_{i=1}^{m}\sum_{j=1}^{n} a_{ij} w_j u_i - \sum_{j=1}^{n}\sum_{i=1}^{m} a_{ij} u_i w_j$$

$$= 0.$$

Therefore $t = c^T w - u^T b$. Since w is primal optimal and u is dual optimal if and only if $c^T w = u^T b$, we are done. ∎

Example. Suppose $x^* = \begin{bmatrix} 1 \\ -1 \\ 2 \\ 0 \\ -3 \end{bmatrix}$ is purported to be an optimal

solution to

$$
\begin{array}{llll}
\min & -4x_1 + 2x_2 - 11x_3 + 4x_4 + 0x_5 & & \\
\text{s.t.} & 2x_1 - x_2 + 3x_3 - 2x_4 + x_5 & = & 6 & \text{tight at } x^* \\
& x_1 + 4x_2 - x_3 - x_4 - 3x_5 & \geq & 4 & \text{tight at } x^* \\
& 4x_1 + 6x_2 - x_3 - 5x_4 + x_5 & \geq & -8 & \text{loose at } x^* \\
\end{array}
$$
$$x_1 \geq 0 \quad x_2 \gtrless 0 \quad x_3 \geq 0 \quad x_4 \geq 0 \quad x_5 \gtrless 0$$

The dual is

$$
\begin{array}{llll}
\max & 6 \cdot y_1 + 4 \cdot y_2 - 8 \cdot y_3 & & \\
\text{s.t.} & 2y_1 + y_2 + 4y_3 & \leq & -4 \\
& -y_1 + 4y_2 + 6y_3 & = & 2 \\
& 3y_1 - y_2 - y_3 & \leq & -11 \\
& -2y_1 - y_2 - 5y_3 & \leq & 4 \\
& y_1 - 3y_2 + y_3 & = & 0 \\
\end{array}
$$
$$y_1 \gtrless 0 \quad y_2 \geq 0 \quad y_3 \geq 0$$

If x^* is primal optimal, complementary slackness insists that if y^* is dual optimal,

$$
\begin{array}{lll}
2y_1^* + y_2^* + 4y_3^* & = & -4 \quad \text{(since } x_1^* > 0\text{)} \\
3y_1^* - y_2^* - y_3^* & = & -11 \quad \text{(since } x_3^* > 0\text{)} \\
y_3^* & = & 0 \quad \text{(since the third primal constraint} \\
& & \text{is loose.)}
\end{array}
$$

The unique solution to this system of three equalities is $y^* = \begin{bmatrix} -3 \\ 2 \\ 0 \end{bmatrix}$
and therefore $\begin{bmatrix} -3 \\ 2 \\ 0 \end{bmatrix}$ must be dual optimal. But $\begin{bmatrix} -3 \\ 2 \\ 0 \end{bmatrix}$ violates
the second and fifth dual constraints, so y^* isn't even dual feasible. It follows that x^* is not primal optimal.

By virtue of complementary slackness, one can sometimes solve a linear program by taking the primal and dual together, and seeking any pair w, u, the first primal feasible and the second dual feasible, together satisfying complementary slackness. Any such pair is primal/dual optimal. In this formulation, no explicit mention of the objective function is made.

3.4 Applications

Farkas' Lemma

The Economic Interpretation of the Dual

Game Theory

3.4.1 Farkas' Lemma

Farkas' Lemma is a remarkably simple characterization of those linear systems that have solutions. Via the Duality Theorem, its proof is trivial.

Suppose one is presented with the problem of finding a vector $x = \begin{bmatrix} x_1 \\ x_2 \end{bmatrix}$ satisfying

$$\begin{aligned} -2x_1 + 3x_2 &\leq -2 \\ 3x_1 - x_2 &\leq 2 \\ -11x_1 - x_2 &\leq -7 \\ x_1 \gtrless 0 \quad x_2 &\gtrless 0. \end{aligned}$$

Let's multiply the first inequality by 2, the second by 5, and the third by 1:

$$
\begin{array}{rcl}
-4x_1 + 6x_2 & \leq & -4 \\
15x_1 - 5x_2 & \leq & 10 \\
-11x_1 - x_2 & \leq & -7 \\
x_1 \gtrless 0 \quad x_2 & \gtrless & 0
\end{array}
$$

and then add them. (Because $2, 5$ and 1 are nonnegative, the senses of the inequalities remain unchanged.) We conclude that $0 \leq -1$. Thus, no x satisfying the inequalities could possibly exist.

More generally, suppose we want to determine if $Ax \leq b$ is solvable. If we can find a nonnegative vector y such that $A^T y = 0$ yet $y^T b < 0$, just as we did in the example, then x cannot exist. The amazing fact is that the converse is true, and both can be proved trivially via the Duality Theorem.

Theorem 28. $Ax \leq b$ has a solution if and only if there is no nonnegative vector y satisfying $A^T y = 0$ and $b^T y < 0$.

Proof. The desired x exists if and only if the linear program P

$$
\begin{array}{rl}
\max & 0^T x \\
& Ax \leq b \\
& x \gtrless 0
\end{array}
$$

is feasible. If P is feasible, it has an optimal value, namely 0. P's dual D is

$$
\begin{array}{rl}
\min & b^T y \\
& A^T y = 0 \\
& y \geq 0
\end{array} \quad .
$$

Since $y = 0$ is feasible in D, D is either unbounded or has an optimal point.

If P is feasible and hence has 0 as its optimal value, then D's optimal value is 0, and hence there can be no nonnegative y satisfying $A^T y = 0$, $b^T y < 0$. On the other hand, if such a y exists, then either D is unbounded, or it has an optimal point of negative value. In either case, P cannot be feasible. ∎

A similar result, proven in the same way, is known as Farkas' Lemma:

Theorem 29. $Ax = b$, $x \geq 0$ has a solution if and only if there is no vector $y \gtrless 0$ satisfying $A^T y \leq 0$ and $b^T y > 0$.

3.4.2 The Economic Interpretation of the Dual.

First, a fact.

Fact 30. If M is an $m \times n$ matrix, the function $f : \mathbb{R}^n \rightarrow \mathbb{R}^m$ taking x to Mx is continuous.

Theorem 31. Consider a primal problem P : $\min c^T x$ in LPS. If
$$Ax = b$$
$$x \geq 0$$
P has at least one nondegenerate basic optimal solution, then there is a positive δ with the following property.

Suppose $t \in \mathbb{R}^m$ and $|t_i| \leq \delta$ for all i. Then

$$P' : \quad \begin{aligned} \min c^T x \\ Ax = b + t \\ x \geq 0 \end{aligned}$$

has an optimal solution and its value is $C^* + (u^*)^T t$, where C^* is the optimal cost of P and u^* is a vector optimal in the dual of P.

Proof. Let $\mathcal{B} = B(1), B(2), \ldots, B(m)$ be an ordered basis corresponding to a nondegenerate optimal bfs w^*, and let B be the matrix of the basic columns in A. Build the Simplex tableau corresponding to $\mathcal{B}$.

- Its entries in rows $1, \ldots, m$ and columns $1, \ldots, n$ are given by $B^{-1}A$,

- column 0 is $B^{-1}b$ (except the $(0,0)$ entry),

- the cost C^* of w^* is $c_B^T B^{-1} b$, and

- row 0 is $c^T - c_B^T B^{-1} A$, as a row vector (except for the $(0,0)$ entry).

$B^{-1}b > 0$. Let ϵ be the smallest component of $B^{-1}b$. Because $f(x) = B^{-1}x$ is continuous, there is a positive δ such that if $|t_i| \leq \delta$ for all i, $B^{-1}(b+t)$ differs from $B^{-1}b$ by at most ϵ in each coordinate (in absolute value). Consequently $B^{-1}(b + t) \geq 0$ and $\mathcal{B}$ yields a basic *feasible* solution in P'. Hence the tableau for $\mathcal{B}$ for P' is the same *everywhere except in column* 0 as the tableau for $\mathcal{B}$ for P. In particular, the 0th row (except for the $(0,0)$ entry) has no negative entries and therefore $\mathcal{B}$ is an optimal basis in P'. In P', the cost of the bfs for $\mathcal{B}$ is $C' = c_B^T B^{-1}(b + t)$. But since $u^* = (B^{-1})^T c_B$ is an optimal solution to P's dual, $C' = (u^*)^T(b + t) = (u^*)^T b + (u^*)^T t = C^* + t^T u^*$. ∎

Is u^* unique? Yes, by complementary slackness.

Think of Q : $\max c^T x$ as a production problem, x_j being the
$\qquad\qquad\qquad Ax \leq b$
$\qquad\qquad\qquad x \geq 0$
output of the jth item, c_j being the profit from the sale of one unit of item j. View the (i,j) entry a_{ij} as the quantity of resource i, e.g., wood, steel or water, needed to produce one unit of item j, and view b_i as the supply of resource i. Q is then just the problem of maximizing profit, subject to the restriction that for each i we cannot produce more items than we have a supply of resource i for. It is not hard to use Theorem 31 to prove that if the equivalent problem $\max c^T x$ has a nondegenerate basic optimal solution,
$\qquad\qquad Ax + Iy = b$
$\qquad\qquad x, y \geq 0$
then, where u^* is optimal in the dual of Q, the optimal value of $\max c^T x$ is $t^T u^*$ larger than the optimal value of Q. This means
$\qquad Ax \leq b + t$
$\qquad x \geq 0$
that u_i^* is the *marginal value* of the ith resource: if the supply of the

ith resource is increased by ϵ (and ϵ is suitably small), the profit will increase by $u_i^* \epsilon$.

3.4.3 Game Theory

Teenagers Rose and Max decide one quiet Saturday night to play a *matrix game*. They write down this 2×2 *payoff matrix*:

$$\begin{array}{c} \qquad\quad Max \\ Rose \begin{array}{|cc|} \hline \$4 & -\$1 \\ \$1 & \$2 \\ \hline \end{array} \end{array}$$

and agree that Max will choose a column and Rose, a row, and then Rose will pay Max the selected amount. For example, if Max chooses 1 and Rose chooses 2, Rose pays Max \$1. If Max chooses 2 and Rose, 1, Rose pays –\$1 (i.e., Max pays \$1). We assume **Rose and Max play optimally.**

This is a $\left\{ \begin{array}{l} \textit{matrix game.} \\ \textit{zero-sum game} \quad - \quad \text{what Rose loses, Max wins, and} \\ \qquad\qquad\qquad\qquad \textit{vice versa.} \end{array} \right.$

How should Max and Rose play? Randomly! Max chooses column 1 with some predetermined probability, and column 2 with the complementary probability. Rose analogously chooses rows with her own selected probabilities. What should those probabilities be?

Suppose Max chooses reals x_1, x_2 and decides to choose column j with probability x_j, $1 \leq j \leq 2$. Of course $x_j \geq 0$, $\sum_{j=1}^{2} x_j = 1$. This vector $x = \begin{bmatrix} x_1 \\ x_2 \end{bmatrix}$ is known as a *mixed strategy*. In our example,

Max will win $4x_1 - x_2 = -1 + 5x_1$ on average, if Rose plays 1.

Max will win $x_1 + 2x_2 = 2 - x_1$ on average, if Rose plays 2.

Max reasons, "If I commit myself to probabilities x_1 and x_2, Rose is going to be able to estimate x_1 and x_2 over many plays of the game. Once she knows x_1 and x_2, she will repeatedly make the same move, the move that minimizes my expected winnings. I'd better choose x_1 so as to *maximize*

$$\min\{-1 + 5x_1, 2 - x_1\};$$

I need the *maximin*." Min$\{-1 + 5x_1, 2 - x_1\}$ is as large as possible when $-1 + 5x_1 = 2 - x_1$ (check this), i.e., $x_1 = \frac{1}{2}$ and $x_2 = \frac{1}{2}$. Then, $-1 + 5x_1 = 2 - x_1 = \frac{3}{2}$ and Max will win \$1.50 on average, per play. So Max decides to choose column 1 exactly half the time.

(If Max had chosen $x_1 = \frac{1}{4}$, for example, min$\{-1 + 5x_1, 2 - x_1\} = \frac{1}{4}$. He'd win only \$0.25 per play.)

And Rose? She reasons, "If I commit myself to probabilities y_1 and y_2, Max is just going to make the same move every time, the move that maximizes his expected winnings. There will be no reason for him to vary his moves. If I choose row i with probability y_i, he will repeatedly choose column 1 if $4y_1 + y_2$, his expected winnings in that case, exceed the $-y_1 + 2y_2$ he can expect to win if he repeatedly chooses column 2. I'd better choose y_1 and y_2 so that max$\{4y_1 + y_2, -y_1 + 2y_2\}$ is as small as possible." So Rows wants to *minimize*

$$\max\{4y_1 + y_2, \ -y_1 + 2y_2\} = \max\{1 + 3y_1, \ 2 - 3y_1\};$$

she wants the *minimax*. This maximum is minimized when $1 + 3y_1 = 2 - 3y_1$ (check it), i.e., $y_1 = \frac{1}{6}$. If Rose sets $y_1 = \frac{1}{6}$, she pays Max \$1.50 on average. (If $y_1 = \frac{1}{2}$, Rose would have to pay Max max$\{1 + \frac{3}{2}, 2 - \frac{3}{2}\} = \$2.50$ on average.) Why \$1.50 again? Duality!

Someday, Rose and Max may tire of this 2×2 game and move up to matrix games played on arbitrary $m \times n$ matrices. To find optimal mixed strategies in a general matrix game, we may assume $A > 0$, as adding a constant to every cell of A cannot affect the choice of mixed strategies. Where e refers to an n- or m-vector of 1's, whichever is appropriate, Max wants to solve

$$
\begin{aligned}
&\text{Find } \alpha \in \mathbb{R}, x \in \mathbb{R}^n \text{ to} \\
&\qquad \max \alpha \\
P_1: \qquad &\qquad \text{s.t. } Ax \ \geq \ \alpha e \\
&\qquad\qquad\ \ e^T x \ = \ 1 \\
&\qquad\qquad\quad\ x \ \geq \ 0
\end{aligned}
$$

while Rose's goal is to solve

$$P_2 : \quad \begin{aligned} \text{Find } \beta &\in \mathbb{R}, y \in \mathbb{R}^m \text{ to} \\ \min\ &\beta \\ \text{s.t.}\quad y^T A &\leq \beta e^T \\ e^T y &= 1 \\ y &\geq 0. \end{aligned}$$

Both P_1 and P_2 are linear programs.
P_1 is equivalent to

$$\begin{aligned} \text{Find } \alpha, x \text{ to max}\quad &\alpha \\ \text{s.t.}\quad A x - \alpha e &\geq 0 \\ -e^T x &= -1 \\ x \geq 0 \quad \alpha &\gtrless 0. \end{aligned}$$

P_2 is equivalent to

$$\begin{aligned} \text{Find } \beta, y \text{ to min}\quad &\beta \\ \text{s.t.}\quad y^T A - \beta e^T &\leq 0 \\ -e^T y &= -1 \\ y \geq 0 \quad \beta &\gtrless 0. \end{aligned}$$

	0	0	0	0	0	−1	
y_1						−1	0
y_2						−1	0
$\vdots$			A			−1	0
						−1	0
y_m						−1	0
β	−1	−1	−1	−1	−1	0	−1
	x_1	x_2	$\cdots$		x_n	α	

Taking $\alpha = 0$ in P_1 and β very large in P_2, we see that both problems are feasible. P_1 and P_2 are duals of each other, and therefore the optimal cost P^* of P_1 equals the optimal cost of P_2. Where $x^*, -P^*$ is optimal in P_1 and $y^*, -P^*$ is optimal in P_2, the pair x^*, y^* is known as a *saddle point*. Neither player wishes to move from it, as any decision to adopt a different mixed strategy only courts disaster. The *value* of the game is defined to be P^*. We have proven

Theorem 32. (The **Minimax Theorem**)

In a matrix game the maximin always equals the minimax.

If Rose and Max follow optimal strategies, each could safely reveal his or her mixed strategy to the other. Knowing Max's optimal mixed strategy will not help Rose, and *vice versa*; in fact, each could calculate the other's optimal strategy.

3.5   Notes

Chvátal's book *Linear Programming* [16] contains an excellent presentation of duality theory. The dual of a linear program is usually just "pulled out of a hat," and one can't help wondering where it came from, why anyone would want to study it, and most importantly, why the Duality Theorem is true. Chvátal motivates study of the dual problem beautifully.

Our proof of the Duality Theorem follows that of Papadimitriou and Steiglitz [44], as does our proof of Corollary 24. Gale [20] credits von Neumann as the first to state the Duality Theorem (in 1947), but his proof was incomplete. Gale, Kuhn and Tucker published the first proof based on von Neumann's notes [21].

Our version of Farkas' Lemma is due to Kuhn [38], the style following Chvátal [16]. Chvátal states the economic importance theorem, leaving the proof as an exercise.

According to Gale [20], the Minimax Theorem was proven originally in 1928—and thus predates the Simplex Algorithm and the Duality Theorem—by von Neumann, whose complicated proof relied on a fixed-point theorem of topology. Later it was shown to be an easy consequence of the Duality Theorem. Our coverage is based on Strang's book [56], which analyzes a simple model of American football: the column player C plays offense, and chooses either to run or to pass, while the defense chooses either to defend against a run or defend against a pass. For additional material on game theory, consult either the book by Owen [43] the one by Shapley [49], those by Shubik [51, 52], or the one by von Neumann and Morgenstern [61].

Chapter 4

The Ellipsoid Algorithm

The first polynomial-time linear programming algorithm, the *Ellipsoid Algorithm* was constructed by Soviet mathematicians, L. G. Khachiyan providing the final details in 1979. It is sometimes known as *Khachiyan's Algorithm* to acknowledge Khachiyan's contribution. The algorithm differs radically from the Simplex Algorithm, in that it almost completely ignores the combinatorial structure of linear programming.

Unfortunately, the Ellipsoid Algorithm does not solve linear programs directly, and understanding it requires some knowledge of affine transformations and, naturally, ellipsoids. Section 4.1 describes the problem the Ellipsoid Algorithm solves directly, and shows how it can be used to solve linear programs. Sections 4.2 and 4.3 provide the necessary background on affine transformations and ellipsoids. The algorithm itself is given in section 4.4.

4.1 LP and Related Problems

Formally, linear programming (LP) is the following. Given an $m \times n$ integral matrix A (of rank m), integral m-vector b, and integral n-vector c, either

73

- Find an n-vector x such that $Ax = b$, $x \geq 0$ and $c^T x$ is minimized over $x \in F = \{ x \mid Ax = b,\ x \geq 0 \}$ (if $F \neq \emptyset$ and the minimum exists), or

- Report infeasibility if $F = \emptyset$, or

- Report unboundedness if $\left\{ c^T x \mid x \in F \right\}$ has no lower bound.

We pose two related questions:

Linear Inequalities (LI): Given an integral $m \times n$ matrix A and an integral m-vector b, find an n-vector x such that $Ax \leq b$, if one exists. If not, say so. (Equality constraints can be written as pairs of inequalities.)

Linear Strict Inequalities (LSI): Given an integral $m \times n$ matrix A and an integral m-vector b, find an x such that $Ax < b$, if one exists. If not, say so.

We will see that the Ellipsoid Algorithm, a polynomial-time algorithm for LSI, can be used to construct a polynomial-time LP algorithm.

Definition. The *size* of an LP $\min c^T x$ is $L = mn + \lceil \log_2 |P| \rceil +$
$$Ax = b$$
$$x \geq 0$$
$n \lceil \log_2 n \rceil$, where P is the product of all nonzero numbers in the problem (in A, b, and c). The sizes of LI and LSI problems are defined similarly.

L approximates the number of bits in the input. The mn term is a lower bound on the number of commas separating the integers in the input. The $\lceil \log_2 |P| \rceil$ is a lower bound on the number of bits needed to represent A, b and c in binary. The $n \lceil \log_2 n \rceil$ appears for a technical reason, but since $m \geq 1$, L is bounded above by the square of $mn + \lceil \log_2 |P| \rceil$, and hence by the square of the number of bits in the input. Any algorithm running in time polynomial in L will run in time polynomial in the number of input bits.

Lemma 33. If x is a basic solution of an integral instance of LPS of size L, then there are integers $D \neq 0$ and $D_1, D_2, ..., D_n$ so that $x_j = D_j / D$ where $|D| < 2^L$ and $|x_j| < 2^L$.

Proof. Let B be the $m \times m$ matrix formed by the basic columns and let $D = \det B$. Each of the $m!$ terms in the formula for the determinant of B, being plus or minus the product of m distinct entries of B, is at most $|P|$ in absolute value. Hence

$$
\begin{aligned}
|D| &\leq m! |P| \\
&\leq n! |P| \\
&\leq 2^{n \lceil \log_2 n \rceil + \log_2 |P|} \\
&< 2^L.
\end{aligned}
$$

An entry x_j of the basic solution x is either (a) $0 = 0/D$ if column j is nonbasic, or (b), applying Cramer's rule, D^{-1} times the determinant of a matrix obtained from B by replacing one column by b. Each term in that determinant is bounded by $|P|$ in absolute value. As above, this determinant can be bounded in absolute value by $m! |P| < 2^L$. Since $D \neq 0$ means that $|D| \geq 1$, $|x_j| < 2^L$. ∎

Lemma 34. Suppose basic feasible solutions w, v of an integral instance of LPS satisfy

$$
|c^T w - c^T v| \leq 2^{-2L}.
$$

Then $c^T w = c^T v$.

Proof. Suppose the lemma is false. Being basic feasible solutions, w and v are rational vectors, each of which has a common denominator less than 2^L; hence $c^T w$ and $c^T v$ are distinct rationals N/D and N'/D', where $|D|, |D'| < 2^L$. Hence

$$
\begin{aligned}
|c^T w - c^T v| &= \left| \frac{N}{D} - \frac{N'}{D'} \right| \\
&= \left| \frac{D'N - DN'}{DD'} \right|.
\end{aligned}
$$

$D'N - DN'$ is a nonzero integer and therefore

$$|c^T w - c^T v| \geq \frac{1}{|D||D'|} > \frac{1}{2^{2L}} = 2^{-2L},$$

a contradiction. ■

Now we prove

Theorem 35. A polynomial-time linear programming algorithm exists if and only if a polynomial-time LI algorithm exists.

Proof. $\Rightarrow$ is easy. Phrase the LI problem as a linear program in standard form with 0 objective function, using Gaussian Elimination to eliminate redundant rows, and solve the LP.

$\Leftarrow$: Let $\min c^T x$ be the LP.
$$Ax = b$$
$$x \geq 0$$

1. Use the LI algorithm to test if the LP is feasible. If not, report infeasibility and halt.

2. Consider the LI system $\left\{ \begin{array}{rcl} Ax & = & b \\ x & \geq & 0 \\ A^T y & \leq & c \\ c^T x - y^T b & = & 0 \end{array} \right.$, a combination of the LP and its dual.

Since the original LP is feasible, this LI system is feasible if and only if the original LP reaches optimality. Run the LI algorithm on this instance of LI, whose size is roughly three times the LP's size. If the LI algorithm reports infeasibility, report unboundedness of LP. Otherwise, if w, u is an optimal pair, w is optimal in the LP instance (and u is optimal in the dual). ■

Theorem 36. The system $Ax \leq b$ has a solution if and only if $Ax < b'$ has a solution, where L is the size of the LI system $Ax \leq b$, $\epsilon = 2^{-2L}$ and $b'_i = b_i + \epsilon$. Furthermore, there is a polynomial-time algorithm which recovers a solution to $Ax \leq b$ from one to $Ax < b'$.

Before we prove this, here is a corollary.

Corollary 37. If there is a polynomial-time algorithm for LSI, there is one for LI (and also for LP).

Proof. To find a vector satisfying $Ax \leq b$, we construct b' with $b'_i = 2^{2L}b_i + 1$ and $A' = 2^{2L}A$, and find a y satisfying the LSI instance $A'x < b'$, if one exists. If so, in polynomial time we construct a solution to $Ax \leq b$ via Theorem 36. The size of $A'x < b'$ is at most twice the square of the size of $Ax \leq b$. ∎

Proof of Theorem 36. We exhibit a method which finds an LI solution given a solution to $Ax < b'$. For $x \in \mathbb{R}^n$, set $\theta_i(x) = A^i x - b_i$.

$$\theta_i(x) \leq 0 \iff x \text{ satisfies the } i\text{th LI constraint.}$$

Let $w \in \mathbb{R}^n$.

Claim. There is a $y \in \mathbb{R}^n$ such that

(1) $\theta_i(y) \leq \max\{0, \theta_i(w)\}$ for $1 \leq i \leq m$ (i.e., no LI constraint satisfied by w is violated by y), and

(2) the set $\{A^i \mid \theta_i(y) \geq 0\}$ spans all of A's rows.

Proof of Claim. It suffices to show that if w does not satisfy (2), we can find a vector w' satisfying (1) for which

$$\{i \mid \theta_i(w') \geq 0\} \underset{\neq}{\supset} \{i \mid \theta_i(w) \geq 0\}.$$

Repeating this at most m times, we obtain a vector y satisfying (1) and (2).

Use θ_i to mean $\theta_i(w)$. Label the constraints so that

$$\begin{aligned} \theta_1, \theta_2, \ldots, \theta_k &\geq 0, \\ \theta_{k+1}, \theta_{k+2}, \ldots, \theta_m &< 0, \end{aligned}$$

and assume (2) does not hold. Then row A^v for some $v > k$ is not a linear combination of $A^1, A^2, \ldots, A^k$, and therefore the system

$$A^i x = 0, \quad 1 \leq i \leq k$$
$$A^v x = 1$$

is solvable. Let u be a solution and consider $w' = w + tu$, where t is the largest real s such that $s(A^i u) + \theta_i \leq 0$ for $i = k + 1, \ldots, m$. (This should remind you of pivoting.) Naturally $0 < t \leq -\theta_v$. By the choice of t,

$$\theta_i(w') = t(A^i u) + \theta_i(w),$$

which is θ_i if $1 \leq i \leq k$. Now

$$\theta_i(w') \leq 0$$

for $k + 1 \leq i \leq m$, but for some $k + 1 \leq i^* \leq m$, $\theta_{i^*}(w') = 0$. This completes the proof of the claim.

Now let $v \in \mathbb{R}^n$ satisfy $A^i v < b_i + \epsilon$ for $i = 1, \ldots, m$. From v we will construct a solution z to the LI system. Replace v by the vector y obtained via the claim (calling the result v). Even now, $A^i v < b_i + \epsilon$ for all i. Label the constraints so that

$A^i v \geq b_i$ for $1 \leq i \leq p$,
$A^i v < b_i$ for $i > p$
$(A^i v < b_i + \epsilon$ for $1 \leq i \leq m)$,
$A^1, A^2, \ldots, A^r$ are independent (so that $r \leq p$), and
$A^1, A^2, \ldots, A^r$ span $A^1, A^2, \ldots, A^m$.

Let z be a solution to the solvable system $A^i x = b_i$, $1 \leq i \leq r$.

Claim. $A^i z \leq b_i$ for *all* $i = 1, 2, \ldots, m$, i.e., z is a solution to the LI system.

Proof of Claim. Let $1 \leq t \leq m$. We know that $A^t = \sum_{i=1}^{r} \lambda_i A^i$ for some $\lambda_i \in \mathbb{R}$.

$$\boxed{\ \underline{\quad\quad} A^t \underline{\quad\quad}\ } = \boxed{\ \lambda_1\ \lambda_2\ \cdots\ \lambda_r\ } \boxed{\begin{array}{c} \underline{\quad} A^1 \underline{\quad} \\ \underline{\quad} A^2 \underline{\quad} \\ \vdots \\ \underline{\quad} A^r \underline{\quad} \end{array}}$$

Equivalently,

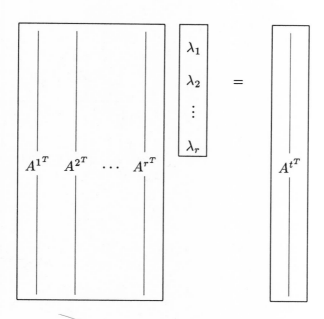

Applying Cramer's rule to a nonsingular $r \times r$ submatrix of A and the corresponding entries in A^t, we conclude that $\lambda_i = \dfrac{D_i}{D}$ where D_i and D are determinants of $r \times r$ submatrices of A and hence are *integers* of absolute value less than 2^L. We may assume $D > 0$ (by negating D_i and D if necessary). Because $\lambda_i = \dfrac{D_i}{D}$,

$$
\begin{aligned}
D[A^t z - b_t] &= \left(\sum_{i=1}^{r} D_i A^i z \right) - D b_t \\
&= \left(\sum_{i=1}^{r} D_i b_i \right) - D b_t
\end{aligned}
$$

by definition of z. Therefore $D[A^t z - b_t]$ is an integer.

Recall that $A^i v < b_i + \epsilon$ for all i and $A^i v \geq b_i$ for $i = 1, \ldots, r$ (since $r \leq p$).

$$D[A^t z - b_t] = -\sum_{i=1}^{r}(-D_i b_i) + \underbrace{\left[\left(-\sum_{i=1}^{r} D_i A^i v\right) + D A^t v\right]}_{=D[-\sum_{i=1}^{r} \lambda_i A^i + A^t]v=0} - D b_t$$

$$= \left[\sum_{i=1}^{r} -D_i(\underbrace{A^i v - b_i}_{0 \leq A^i v - b_i < \epsilon})\right] + D(\underbrace{A^t v - b_t}_{<\epsilon})$$

But $D[A^t v - b_t] < D\epsilon$ and $|-D_i(\underbrace{A^i v - b_i}_{\geq 0})| = |D_i|(\underbrace{A^i v - b_i}_{\geq 0}) < |D_i|\epsilon$.

So

$$
\begin{aligned}
D[A^t z - b_t] &< \left[\sum_{i=1}^{r} |D_i| + D\right] \epsilon \\
&< \left[\left(\sum_{i=1}^{r} 2^L\right) + 2^L\right] \epsilon \\
&\leq (m+1)2^L 2^{-2L} \\
&< 1.
\end{aligned}
$$

Therefore $D[A^t z - b_t]$, being an integer, is nonpositive. Therefore $A^t z \leq b_t$, and the proofs of the claim and the theorem are complete. ∎

4.2   Affine Transformations and Ellipsoids

Definition. The map $T(x) = t + Qx$ is an *affine transformation* if $t \in \mathbb{R}^n$ and Q is an $n \times n$ nonsingular matrix.

If T is an affine transformation, then T's inverse $T^{-1}(x) = Q^{-1}(x - t)$ is also an affine transformation.

Lemma 38. If T is invertible and $A, B \subseteq \mathbb{R}^n$, $T(A \cap B) = T(A) \cap T(B)$.

Proof. Exercise.   ■

Recall that the *length* of a vector x is $\|x\| = \sqrt{x^T x}$. Multiplication by an orthogonal matrix U preserves length: $\|Ux\| = \|x\|$ for all x.

Definition. For $v \in \mathbb{R}^n$ and $r > 0$,

$$S(v, r) = \{\, x \in \mathbb{R}^n \mid \|x - v\| \le r \,\},$$

the sphere of radius r about v. (The n is implicit.)

Definition. Where T is an affine transformation, the image of the unit sphere $S(0, 1)$ is an *ellipsoid*. Equivalently, if $T(x) = t + Qx$, $T(S(0, 1)) = \{\, t + Qx \mid x^T x \le 1 \,\}$.

If $y = t + Qx$, $x = Q^{-1}(y - t)$. Thus

$$
\begin{aligned}
T(S(0,1)) &= \left\{\, y \mid (Q^{-1}(y - t))^T (Q^{-1}(y - t)) \le 1 \,\right\} \\
&= \left\{\, y \mid (y - t)^T Q^{-1^T} Q^{-1} (y - t) \le 1 \,\right\} \\
&= \left\{\, y \mid (y - t)^T B^{-1} (y - t) \le 1 \,\right\},
\end{aligned}
$$

where $B = QQ^T$. Since Q is nonsingular, B is positive definite. Conversely, if B is positive definite, $B = QQ^T$ for some nonsingular Q. Therefore $\{\, x \mid \underbrace{(x - t)^T B^{-1} (x - t)}_{\text{a ``quadratic form''}} \le 1 \,\}$ is an ellipsoid, where

$t \in \mathbb{R}^n$ and B is positive definite, and every ellipsoid can be put in this form.

Let
$$E = \{x \in \mathbb{R}^n | x^T B^{-1} x \leq 1\}$$
for a positive definite $n \times n$ B. By the Spectral Theorem (Theorem 4 of chapter 1),
$$B = U^{-1} \Lambda U$$

where Λ is a diagonal matrix whose jth diagonal entry is the jth eigenvalue λ_j of B and where U is the orthogonal matrix whose jth row is the transpose of the normalized eigenvector corresponding to λ_j. If D is a diagonal matrix whose diagonal entries d_j are non-negative, define $\sqrt{D}$ to be the diagonal matrix with $\sqrt{d_j}$ in the jth diagonal cell; clearly $\sqrt{D}\sqrt{D} = D$. Thus
$$B = (\sqrt{\Lambda}U)^T(\sqrt{\Lambda}U)$$
—recall that $U^{-1} = U^T$—and therefore
$$
\begin{aligned}
B^{-1} &= U^T(\sqrt{\Lambda})^{-1}(\sqrt{\Lambda})^{-1}U \\
&= (\sqrt{\Lambda^{-1}}U)^T(\sqrt{\Lambda^{-1}}U).
\end{aligned}
$$

In other words,
$$
\begin{aligned}
E &= \{x \in \mathbb{R}^n | x^T(\sqrt{\Lambda^{-1}}U)^T(\sqrt{\Lambda^{-1}}U)x \leq 1\} \\
&= \{x \in \mathbb{R}^n | [(\sqrt{\Lambda^{-1}}U)x]^T[(\sqrt{\Lambda^{-1}}U)x] \leq 1\}.
\end{aligned}
$$

Now let $y = (\sqrt{\Lambda^{-1}}U)x$, so that $x = (U^T\sqrt{\Lambda})y$. Then
$$E = \{(U^T\sqrt{\Lambda})y | y^T y \leq 1\}.$$

Starting from the unit sphere $S(0,1)$, the $\sqrt{\Lambda}$ "stretches" the co-ordinates, expanding the jth by a factor of $\sqrt{\lambda_j}$, while the U^T rotates the figure. Thus an ellipsoid $E = \{x | x^T B^{-1} x \leq 1\}$ can be obtained from $S(0,1)$ by stretching the jth axis by a factor of $\sqrt{\lambda_j}$, $j = 1, 2, ..., n$, and then rotating the resulting figure. Replacing $E = \{x | x^T B^{-1} x \leq 1\}$ by $E = \{x | (x-t)^T B^{-1}(x-t) \leq 1\}$ does nothing more than translate the center of E from the origin to t.

Example. Let $B = \begin{bmatrix} 5 & 2 \\ 2 & 8 \end{bmatrix}$, whose inverse $B^{-1} = \begin{bmatrix} \frac{2}{9} & -\frac{1}{18} \\ -\frac{1}{18} & \frac{5}{36} \end{bmatrix}$.

Let

$$
\begin{aligned}
E &= \{x \in \mathbb{R}^2 | x^T B^{-1} x \le 1\} \\
&= \{x | \frac{2}{9} x_1^2 - \frac{1}{9} x_1 x_2 + \frac{5}{36} x_2^2 \le 1\}.
\end{aligned}
$$

Corresponding to the eigenvalues 4 and 9 of B are the normalized eigenvectors $\begin{bmatrix} \frac{2}{\sqrt{5}} \\ -\frac{1}{\sqrt{5}} \end{bmatrix}$ and $\begin{bmatrix} \frac{1}{\sqrt{5}} \\ \frac{2}{\sqrt{5}} \end{bmatrix}$, respectively. Therefore

$$
B = \begin{bmatrix} \frac{2}{\sqrt{5}} & \frac{1}{\sqrt{5}} \\ -\frac{1}{\sqrt{5}} & \frac{2}{\sqrt{5}} \end{bmatrix} \begin{bmatrix} 4 & 0 \\ 0 & 9 \end{bmatrix} \begin{bmatrix} \frac{2}{\sqrt{5}} & -\frac{1}{\sqrt{5}} \\ \frac{1}{\sqrt{5}} & \frac{2}{\sqrt{5}} \end{bmatrix}
$$

or

$$
B = \left(\begin{bmatrix} \frac{2}{\sqrt{5}} & \frac{1}{\sqrt{5}} \\ -\frac{1}{\sqrt{5}} & \frac{2}{\sqrt{5}} \end{bmatrix} \begin{bmatrix} 2 & 0 \\ 0 & 3 \end{bmatrix} \right) \left(\begin{bmatrix} 2 & 0 \\ 0 & 3 \end{bmatrix} \begin{bmatrix} \frac{2}{\sqrt{5}} & -\frac{1}{\sqrt{5}} \\ \frac{1}{\sqrt{5}} & \frac{2}{\sqrt{5}} \end{bmatrix} \right).
$$

Thus E can be obtained from $S(0,1)$ by doubling the first coordinate and tripling the second, and then rotating the figure via the orthogonal matrix $\begin{bmatrix} \frac{2}{\sqrt{5}} & \frac{1}{\sqrt{5}} \\ -\frac{1}{\sqrt{5}} & \frac{2}{\sqrt{5}} \end{bmatrix}$. Here is E:

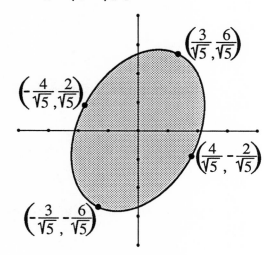

4.3 Basic Lemmas

We need the following series of lemmas and definitions.

Lemma 39. (See Rudin [46].)

Suppose that $P \subseteq \mathbb{R}^n$ has volume μ. Let $T(x) = t + Qx$, where Q is an arbitrary $n \times n$ matrix. Then $T(P)$ has volume $|\det Q|\mu$.

Lemma 40. Given $a \in \mathbb{R}^n$, there is an orthogonal matrix U such that $Ua = [\|a\| \ 0 \ 0 \ \cdots \ 0]^T$.

Proof. It suffices to consider unit-length vectors $a \in \mathbb{R}^n$ that have at least two nonzero entries. We will construct an orthogonal U for which Ua has strictly fewer nonzero entries than a. By multiplying together matrices of this form, we can reduce the number of nonzeroes in a to one. That entry will be ± 1, since multiplication by orthogonal matrices leaves lengths unchanged. Multiplying by a permutation matrix or its negative will convert the result to $[1 \ 0 \ 0 \ \cdots \ 0]^T$.

Suppose $a_k \neq 0$ and $a_j \neq 0$, $k < j$. Let U be the matrix obtained from the $n \times n$ identity matrix by replacing the $\begin{bmatrix} 1 & 0 \\ 0 & 1 \end{bmatrix}$ in positions (k,k), (k,j), (j,k) and (j,j) by

$$\frac{1}{\sqrt{a_k^2 + a_j^2}} \begin{bmatrix} a_k & a_j \\ a_j & -a_k \end{bmatrix}.$$

It is easy to verify that U is orthogonal. Moreover, if $l \neq k,j$, the lth component of Ua is a_l. Its jth entry is

$$\frac{a_j a_k - a_k a_j}{\sqrt{a_k^2 + a_j^2}} = 0,$$

and the proof is complete. ∎

The Gram-Schmidt orthonormalization process can be used to give an alternate proof of Lemma 40.

Definition. For P a polytope in $\mathbb{R}^n$, its *interior* int $P = \{\, x \mid \exists \epsilon > 0 \text{ such that } S(x,\epsilon) \subseteq P\}$.

Definition. If $T \subseteq \mathbb{R}^n$, the *convex hull* $C(T)$ of T is the intersection of all convex sets containing T.

Lemma 41. If int$(P) \neq \emptyset$, P has $n+1$ vertices whose convex hull has positive volume.

Proof. By the fact that every point in a polytope $P \subseteq \mathbb{R}^n$ can be written as the convex combination of at most $n+1$ vertices (Lemma 12), and the fact that P is convex, we can write P as the union of $C(T)$ over all sets T containing $n+1$ vertices. Choose x in int(P) and $\epsilon > 0$ so that $S(x,\epsilon) \subseteq P$. Then $S(x,\epsilon)$ is the union, over sets T containing $n+1$ vertices, of $S(x,\epsilon) \cap C(T)$. Since $S(x,\epsilon)$ has positive volume, so does one $S(x,\epsilon) \cap C(T)$, and therefore so does one $C(T)$. ∎

Definition. Let $\Gamma_n = \{x \in \mathbb{R}^n \mid x \geq 0, \sum x_j \leq 1\}$.

Lemma 42. The n-dimensional volume of Γ_n is $1/n!$.

Proof. Exercise. Use induction, integration, and the fact that a cross section of Γ_n is a "scaled-down" version of Γ_{n-1}.

Lemma 43. The volume of the convex hull of points $v_0, v_1, v_2, \ldots, v_n \in \mathbb{R}^n$ is at least

$$\frac{1}{n!} \left| \det \begin{bmatrix} 1 & 1 & 1 & & 1 \\ | & | & | & & | \\ v_0 & v_1 & v_2 & \cdots & v_n \\ | & | & | & & | \end{bmatrix} \right|.$$

Proof. Fix n. Define n-vectors $z_j = v_j - v_0$ for $j = 1, 2, \ldots, n$. The volume of the convex hull of $v_0, v_1, v_2, \ldots, v_n$ is clearly the same as that of the convex hull C of $z_0, z_1, z_2, \ldots, z_n$. Define $n \times n$ matrix Z

by putting z_j in column j for $j = 1, 2, ..., n$ (so $Z_j = z_j$). It is easy to see (by subtracting column one from every other column) that

$$\det Z = \det \begin{bmatrix} 1 & 1 & 1 & & 1 \\ | & | & | & & | \\ v_0 & v_1 & v_2 & \cdots & v_n \\ | & | & | & & | \end{bmatrix}.$$

Let T be the mapping $x \mapsto Zx$. By Lemmas 39 and 42, the volume of $T(\Gamma_n)$ is $\frac{1}{n!}|\det Z|$. Showing that $T(\Gamma_n) \subseteq C$ will complete the proof.

Let $e_0 = 0 \in \mathbb{R}^n$ and for $j \geq 1$, let $e_j \in \mathbb{R}^n$ be the zero vector, except for a one in the jth position. Let $x = [x_1 \ \ x_2 \ \ \cdots \ \ x_n]^T$ be an arbitrary point in Γ_n and let $x_0 = 1 - x_1 - x_2 - \cdots - x_n$; $x = \sum_{j=0}^{n} x_j e_j$. Now $T(x) = Zx = \sum_{j=0}^{n} x_j Z e_j = \sum_{j=0}^{n} x_j z_j$, a convex combination of $z_0, z_1, ..., z_n$. Because a convex combination of points in a convex set is itself in the set, $T(x)$ is in C. Since x was an arbitrary point of Γ_n, it follows that $T(\Gamma_n) \subseteq C$ and the proof is complete. ∎

In fact, the volume of the convex hull is exactly the quantity given in Lemma 43, but we will need only the lower bound.

4.4   The Algorithm

The Ellipsoid Algorithm solves LSI in polynomial time. In each iteration, it produces an ellipsoid E which contains the feasible region $F = \{ x \mid Ax < b \}$, and tests the center t of E for feasibility. If $t \in F$, it prints t and halts, having successfully solved the problem. Otherwise, it finds an LSI constraint $A^i x < b_i$ that is violated by t

(i.e., $A^i t \geq b_i$).

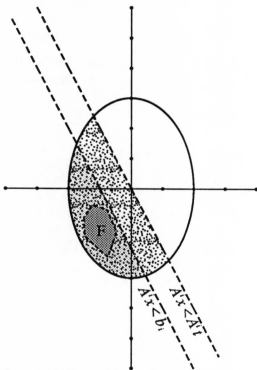

Because $A^i t \geq b_i$, no feasible points can lie in the half ellipsoid "above" the hyperplane passing through t that is parallel to the violated constraint. (This is the line passing through the origin in the picture.) The other half of E (shaded in the picture) contains all of F. The idea is to build a new ellipsoid E' which contains that half-ellipsoid and *whose volume is significantly smaller than the volume of E. E'* will be the ellipsoid for the next iteration.

In this way, the Ellipsoid Algorithm generates a sequence of ellipsoids all of which contain F and which shrink in volume. If any of the ellipsoids has a feasible center, the Ellipsoid Algorithm prints that point and halts.

But what if F is empty? None of the centers of the shrinking ellipsoids is feasible, of course, and the algorithm could run forever.

We shall see, however, that in this case the volume of the feasible region can be bounded below by a positive ν. If so many iterations have been performed that the current ellipsoid has volume less than ν, the algorithm can safely print "$F = \emptyset$" and halt.

(In fact, the ellipsoids may not contain all of F, if F is enormous, just a "large enough" fraction of it. This is a minor technicality.)

The Ellipsoid Algorithm

1. Let $l = 0$, and let $K = 16n(n + 1)L$.
   Let t_0 be the 0 n-vector, and let $B_0 = n^2 2^{2L} I$.
   (The lth ellipsoid is $E_l = \left\{ x \mid (x - t_l)^T B_l^{-1}(x - t_l) \leq 1 \right\}$, where B_l is positive definite. The initial ellipsoid is $S(0, n2^L)$.)
2. For $l = 0$ to K do
   If t_l satisfies $Ax < b$, print t_l and halt. Otherwise, build a new, smaller ellipsoid:
   Find an i such that $A^i t_l \geq b_i$. (Thus $A^i x < b_i$ is violated by t_l.)
   Let $a = (A^i)^T$.
   Let $t_{l+1} = t_l - \dfrac{1}{n + 1} \dfrac{B_l a}{\sqrt{a^T B_l a}}$.
   Let $B_{l+1} = \dfrac{n^2}{n^2 - 1} \left[B_l - \underbrace{\dfrac{2}{n + 1} \dfrac{(B_l a)(B_l a)^T}{a^T B_l a}}_{n \times n \text{ matrix}} \right].$
3. Print "infeasible" and halt; after K iterations, the ellipsoid is too small for the LSI instance to be feasible.

End.

Example. Suppose $B_l = \begin{bmatrix} 5 & 2 \\ 2 & 8 \end{bmatrix}$, $t_l = \begin{bmatrix} 0 \\ 0 \end{bmatrix}$, and among the LSI constraints is $x_1 - x_2 < -1$. From

$$B_l a = \begin{bmatrix} 5 & 2 \\ 2 & 8 \end{bmatrix} \begin{bmatrix} 1 \\ -1 \end{bmatrix} = \begin{bmatrix} 3 \\ -6 \end{bmatrix}$$

and $a^T B_l a = 9$, we infer that

$$t_{l+1} = \begin{bmatrix} -\frac{1}{3} \\ \frac{2}{3} \end{bmatrix}, \qquad B_{l+1} = \begin{bmatrix} \frac{52}{9} & \frac{40}{9} \\ \frac{40}{9} & \frac{64}{9} \end{bmatrix}.$$

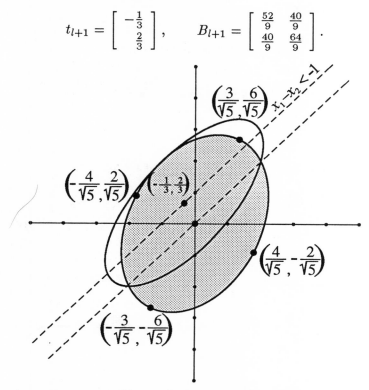

For the new ellipsoid, we could use any ellipsoid that contains the part of the shaded region which lies above the line $x_1 - x_2 = -1$. Points outside of that region are necessarily infeasible. We choose to include the half ellipsoid above the line $x_1 - x_2 = 0$ (which is parallel to the violated constraint and passes through the ellipsoid's center), since the formula describing it is comparatively simple.

As usual, we will assume exact real arithmetic. Each iteration takes time bounded by a polynomial in L, and since K likewise is bounded by a polynomial in L, the overall time expended is polynomial. But is the algorithm correct? To prove the Ellipsoid Algorithm correct we prove

Theorem 44. Let B_l be positive definite, let $t_l \in \mathbb{R}^n$, and let

$a \in \mathbb{R}^n$ be nonzero. Construct B_{l+1} and t_{l+1} as in step 2 of the Ellipsoid Algorithm. Then

(a) B_{l+1} is positive definite (so that

$$E_{l+1} = \left\{ x \mid (x - t_{l+1})^T B_{l+1}^{-1} (x - t_{l+1}) \leq 1 \right\}$$

is an ellipsoid).

(b) Let $H E_l[a]$ be the half-ellipsoid containing F:

$$
\begin{aligned}
H E_l[a] \quad &= \quad E_l \cap \{ x \in \mathbb{R}^n \mid a^T (x - t_l) \leq 0 \} \\
&= \quad \{ x \in \mathbb{R}^n \mid (x - t_l)^T B_l^{-1} (x - t_l) \leq 1 \text{ and} \\
&\qquad a^T (x - t_l) \leq 0 \}.
\end{aligned}
$$

Then $H E_l[a] \subseteq E_{l+1}$.

(c) $\dfrac{\text{volume}(E_{l+1})}{\text{volume}(E_l)} < 2^{\frac{-1}{2(n+1)}}$ (so that E_{l+1} is "significantly smaller" than E_l).

Let us first prove our result in a simple special case.

Lemma 45.   Let $n \geq 2$ and let

$$E = \left\{ x \in \mathbb{R}^n \mid (x - t)^T B^{-1} (x - t) \leq 1 \right\},$$

where

$$t = \begin{bmatrix} \frac{-1}{n+1} \\ 0 \\ 0 \\ \vdots \\ 0 \end{bmatrix}$$

and

$$B = \begin{bmatrix} \dfrac{n^2}{(n+1)^2} & 0 & 0 & 0 & 0 \\ 0 & \dfrac{n^2}{n^2-1} & 0 & 0 & 0 \\ 0 & 0 & \dfrac{n^2}{n^2-1} & 0 & 0 \\ 0 & 0 & 0 & \ddots & 0 \\ 0 & 0 & 0 & 0 & \dfrac{n^2}{n^2-1} \end{bmatrix}.$$

Then

(a) B is positive definite (so E is an ellipsoid).

(b) The hemisphere $HS = \left\{ x \mid x^T x \leq 1 \text{ and } x_1 \leq 0 \right\} \subseteq E$.

(c) $\dfrac{\text{volume}(E)}{\text{volume}(S(0,1))} < 2^{\frac{-1}{2(n+1)}}$.

Proof.

(a) $B = QQ^T$ where

$$Q = \begin{bmatrix} \dfrac{n}{n+1} & 0 & 0 & 0 & 0 \\ 0 & \dfrac{n}{\sqrt{n^2-1}} & 0 & 0 & 0 \\ 0 & 0 & \dfrac{n}{\sqrt{n^2-1}} & 0 & 0 \\ 0 & 0 & 0 & \ddots & 0 \\ 0 & 0 & 0 & 0 & \dfrac{n}{\sqrt{n^2-1}} \end{bmatrix}.$$

(b) Let $x \in HS$.
$$(x - t)^T B^{-1} (x - t)$$

$$= \frac{(n+1)^2}{n^2} \left(x_1 + \frac{1}{n+1} \right)^2 + \frac{n^2-1}{n^2} \sum_{i=2}^{n} x_i^2$$

$$= \left[\frac{n^2 - 1}{n^2}\right] x^T x + \frac{2n+2}{n^2} x_1^2 + \frac{2n+2}{n^2} x_1 + \frac{1}{n^2}$$

$$\leq 1 + \frac{2n+2}{n^2}(x_1^2 + x_1),$$

since $x \in S(0,1)$. Because $x \in HS$, $x_1^2 + x_1 = x_1(x_1 + 1) \leq 0$. Hence $(x - t)^T B^{-1}(x - t) \leq 1$ and therefore $HS \subseteq E$.

(c) $E = T(S(0,1))$ where $T(x) = t + Qx$. By Lemma 39,

$$\text{volume}(E) = |\det Q|[\text{volume}(S(0,1))].$$

Det $Q = \frac{n}{n+1}\left(\frac{n^2}{n^2 - 1}\right)^{\frac{n-1}{2}}$.   For all real y, $1 + y \leq e^y$. Therefore

$$\frac{n}{n+1} = 1 - \frac{1}{n+1} \leq e^{-1/(n+1)}$$

and

$$\frac{n^2}{n^2 - 1} = 1 + \frac{1}{n^2 - 1} \leq e^{1/(n^2 - 1)}.$$

It follows that

$$|\det Q| \leq e^{-\frac{1}{n+1} + \frac{n-1}{2(n^2-1)}} = e^{-\frac{1}{2(n+1)}} < 2^{-\frac{1}{2(n+1)}}. \quad \blacksquare$$

The next lemma relates the special case of Lemma 45 to the general case of Theorem 44.

Lemma 46. Let B_l be positive definite, $t_l \in \mathbb{R}^n$, and let $a \in \mathbb{R}^n$ be nonzero. Let $E_l = \left\{ x \mid (x - t_l)^T B_l^{-1}(x - t_l) \leq 1 \right\}$ and let $HE_l[a] = E_l \cap \left\{ x \mid a^T(x - t_l) \leq 0 \right\}$, as in the statement of Theorem 44. Let t_{l+1} and B_{l+1} be obtained via step 2 of the Ellipsoid Algorithm. Let HS, E, and t be as in the previous lemma. Then there is an affine transformation T such that:

(a) The unit sphere maps onto E_l:

$$T(S(0,1)) = E_l = \left\{ x \mid (x - t_l)B_l^{-1}(x - t_l) \leq 1 \right\}.$$

(b) E maps onto E_{l+1}:

$$T(E) = E_{l+1} = \left\{ x \mid (x - t_{l+1})^T B_{l+1}^{-1} (x - t_{l+1}) \le 1 \right\}.$$

(c) The hemisphere maps onto the half ellipsoid:

$$T(HS) = HE_l[a].$$

Proof. Because B_l is positive definite, $B_l = QQ^T$ for some nonsingular Q. Lemma 40 implies that there is an orthogonal matrix U^T such

that $U^T(Q^T a) = \begin{bmatrix} \|Q^T a\| \\ 0 \\ 0 \\ \vdots \\ 0 \end{bmatrix}$. Let T be the map $T(x) = t_l + (QU)x$,

an affine transformation.

(a)

$$\begin{aligned} T(S(0,1)) &= \left\{ T(x) \mid x^T x \le 1 \right\} \\ &= \left\{ y \mid (T^{-1}y)^T (T^{-1}y) \le 1 \right\} \\ &= \left\{ y \mid (y - t_l)^T (Q^{-1})^T \underbrace{UU^T}_{I} Q^{-1}(y - t_l) \le 1 \right\}. \end{aligned}$$

$(Q^{-1})^T Q^{-1} = B_l^{-1}$ implies that

$$T(S(0,1)) = \left\{ y \mid (y - t_l)^T B_l^{-1}(y - t_l) \le 1 \right\} = E_l.$$

(b)

$$\begin{aligned} B_{l+1} &= \frac{n^2}{n^2 - 1} \left[B_l - \frac{2}{n+1} \frac{(B_l a)(B_l a)^T}{a^T B_l a} \right] \\ &= \frac{n^2}{n^2 - 1} \left[B_l - \frac{2}{n+1} \frac{(QUU^T Q^T a)(a^T QUU^T Q^T)}{a^T QQ^T a} \right]. \end{aligned}$$

But $U^T Q^T a = \begin{bmatrix} \|Q^T a\| \\ 0 \\ 0 \\ \vdots \\ 0 \end{bmatrix}$ implies that

$$(U^T Q^T a)(a^T Q U) = \begin{bmatrix} \|Q^T a\|^2 & 0 & 0 & 0 & 0 \\ 0 & 0 & 0 & 0 & 0 \\ 0 & 0 & 0 & 0 & 0 \\ 0 & 0 & 0 & \ddots & 0 \\ 0 & 0 & 0 & 0 & 0 \end{bmatrix}.$$

So

$$\frac{n^2 - 1}{n^2} B_{l+1}$$

$$= B_l - \frac{1}{\|Q^T a\|^2} \left[\frac{2}{n+1} QU \begin{bmatrix} \|Q^T a\|^2 & 0 & 0 & 0 & 0 \\ 0 & 0 & 0 & 0 & 0 \\ 0 & 0 & 0 & 0 & 0 \\ 0 & 0 & 0 & \ddots & 0 \\ 0 & 0 & 0 & 0 & 0 \end{bmatrix} U^T Q^T \right]$$

$$= QQ^T - \frac{2}{n+1} QU \begin{bmatrix} 1 & 0 & 0 & 0 & 0 \\ 0 & 0 & 0 & 0 & 0 \\ 0 & 0 & 0 & 0 & 0 \\ 0 & 0 & 0 & \ddots & 0 \\ 0 & 0 & 0 & 0 & 0 \end{bmatrix} U^T Q^T$$

$$= QU \left[I - \frac{2}{n+1} \begin{bmatrix} 1 & 0 & 0 & 0 & 0 \\ 0 & 0 & 0 & 0 & 0 \\ 0 & 0 & 0 & 0 & 0 \\ 0 & 0 & 0 & \ddots & 0 \\ 0 & 0 & 0 & 0 & 0 \end{bmatrix} \right] U^T Q^T$$

$$= QU \begin{bmatrix} \dfrac{n-1}{n+1} & 0 & 0 & 0 & 0 \\ 0 & 1 & 0 & 0 & 0 \\ 0 & 0 & 1 & 0 & 0 \\ 0 & 0 & 0 & \ddots & 0 \\ 0 & 0 & 0 & 0 & 1 \end{bmatrix} U^T Q^T.$$

Hence $B_{l+1} = QUBU^TQ^T$ where B is as in Lemma 45. Also,

$$x - t_{l+1} = x - t_l + \frac{QUU^TQ^Ta}{(n+1)\sqrt{a^TQQ^Ta}}.$$

But

$$(U^TQ^T)a = \begin{bmatrix} \|Q^Ta\| \\ 0 \\ 0 \\ \vdots \\ 0 \end{bmatrix}.$$

So

$$\begin{aligned}
x - t_{l+1} &= x - t_l + \frac{QU \begin{bmatrix} \|Q^Ta\| \\ 0 \\ 0 \\ \vdots \\ 0 \end{bmatrix}}{(n+1)\|Q^Ta\|} \\
&= x - t_l + QU \begin{bmatrix} \dfrac{1}{n+1} \\ 0 \\ 0 \\ \vdots \\ 0 \end{bmatrix} \\
&= QU(U^{-1}Q^{-1}(x - t_l) - t) \\
&= QU[T^{-1}(x) - t].
\end{aligned}$$

Therefore

$$\begin{aligned}
T(E) &= \left\{ T(x) \mid (x - t)^T B^{-1}(x - t) \leq 1 \right\} \\
&= \left\{ x \mid (T^{-1}(x) - t)^T B^{-1}(T^{-1}(x) - t) \leq 1 \right\} \\
&= \left\{ x \mid (x - t_{l+1})^T (Q^{-1})^T U B^{-1} U^T Q^{-1}(x - t_{l+1}) \leq 1 \right\}.
\end{aligned}$$

But, as we showed above, $B_{l+1}^{-1} = (Q^{-1})^T U B^{-1} U^T Q^{-1}$, and therefore

$$T(E) = \left\{ x \mid (x - t_{l+1})^T B_{l+1}^{-1}(x - t_{l+1}) \leq 1 \right\}.$$

This completes the proof of (b).

(c) Point y is in $T(\{\, x \in \mathbb{R}^n \mid x_1 \leq 0 \,\})$ if and only if $y = t_l + QUx$ for some $x \in \mathbb{R}^n$ with $x_1 \leq 0$.

$$x_1 \leq 0 \iff \|Q^T a\| x_1 \leq 0 \iff x^T \begin{bmatrix} \|Q^T a\| \\ 0 \\ 0 \\ \vdots \\ 0 \end{bmatrix} \leq 0$$

$$\iff x^T U^T Q^T a \leq 0$$

Therefore

$$y \in T(\{\, x \mid x_1 \leq 0 \,\}) \iff \underbrace{[U^{-1}Q^{-1}(y - t_l)]}_{x}{}^T U^T Q^T a \leq 0$$

$$\iff (y - t_l)^T a \leq 0$$
$$\iff a^T (y - t_l) \leq 0.$$

So $T(\{\, x \mid x_1 \leq 0 \,\}) = \{\, y \mid a^T(y - t_l) \leq 0 \,\}$.

By Lemma 38,

$$\begin{aligned} T(HS) &= T(S(0,1) \cap \{\, x \mid x_1 \leq 0 \,\}) \\ &= T(S(0,1)) \cap \{\, x \mid a^T(x - t_l) \leq 0 \,\} \\ &= E_l \cap \{\, x \mid a^T(x - t_l) \leq 0 \,\} \\ &= HE_l[a]. \quad \blacksquare \end{aligned}$$

Now we complete the proof of Theorem 44.

Proof of Theorem 44.

(a) By the proof of Lemma 46(b),

$$\begin{aligned} B_{l+1} &= QUBU^T Q^T \\ &= (QU\sqrt{B})(QU\sqrt{B})^T \end{aligned}$$

and $QU\sqrt{B}$ is nonsingular, so B_{l+1} is positive definite and $E_{l+1} = \{\, x \mid (x - t_{l+1})^T B_{l+1}^{-1}(x - t_{l+1}) \leq 1 \,\}$ is an ellipsoid.

(b) $HE_l[a] = T(HS)$ by part (c) of Lemma 46, and by Lemma 45(b), $HS \subseteq E$. Hence $HE_l[a] \subseteq \underbrace{T(E) = E_{l+1}}_{\text{Lemma 46(b)}}$.

(c) By Lemma 39,

$$
\begin{aligned}
\frac{\text{volume}(E_{l+1})}{\text{volume}(E_l)} &= \frac{\text{volume}(T(E))}{\text{volume}(T(S(0,1)))} \\
&= \frac{|\det QU|(\text{volume}(E))}{|\det QU|(\text{volume}(S(0,1)))} \\
&= \frac{\text{volume}(E)}{\text{volume}(S(0,1))} \\
&< 2^{-\frac{1}{2(n+1)}} \text{ by Lemma 45(c).} \quad \blacksquare
\end{aligned}
$$

One more lemma. If any ellipsoid has volume less than $2^{-(n+2)L}$, the LSI system must be infeasible:

Lemma 47. If an LSI system of size L has a solution, the set of solutions within the sphere $S(0, n2^L) = E_0$ has volume at least $2^{-(n+2)L}$.

Proof. We state a claim.

Claim. If $Ax < b$ has a solution z and L is the system's size, then

$$
\begin{aligned}
Ax &< b \\
x_j &< 2^L \quad \forall j \\
x_j &> -2^L \quad \forall j
\end{aligned}
$$

has a solution.

Proof Sketch. Whenever $Ax < b$ has a solution, so does $Ax \leq b$. The system $Ax \leq b$ can be shown to have a solution v with $|v_j| < 2^L$ for all j by analyzing the equivalent system $Ax + Iy = b$, $y \geq 0$. For a sufficiently small $\epsilon > 0$, $v + \epsilon(z - v)$ satisfies all the strict inequalities.

Hence the polytope $\begin{array}{rcl} Ax & \leq & b \\ x_j & \leq & 2^L \quad \forall j \\ x_j & \geq & -2^L \quad \forall j \end{array}$ has an interior point. By
Lemma 41, it has $n + 1$ vertices $v_0, v_1, \ldots, v_n$ the convex hull C of
which has positive volume. All interior points of C are solutions to
$Ax < b$ within the sphere $S(0, n2^L)$.

Lemma 43 tells us that C's n-dimensional volume is at least

$$\frac{1}{n!} \left| \det \begin{bmatrix} 1 & 1 & & 1 \\ | & | & & | \\ v_0 & v_1 & \cdots & v_n \\ | & | & & | \end{bmatrix} \right| > 0.$$

Each v_l, being a vertex of $\begin{array}{rcl} Ax & \leq & b \\ x_j & \leq & 2^L \quad \forall j \\ x_j & \geq & -2^L \quad \forall j, \end{array}$ can be written as $\dfrac{u_l}{D_l}$
where u_l is an integral vector and D_l is a determinant of absolute
value less than 2^L (check this). Thus C's volume is at least

$$\frac{1}{n!} \left| \det \begin{bmatrix} D_0 & D_1 & & D_n \\ | & | & & | \\ u_0 & u_1 & \cdots & u_n \\ | & | & & | \end{bmatrix} \right| \frac{1}{|D_0||D_1| \cdots |D_n|}.$$

$\text{Det} \begin{bmatrix} D_0 & D_1 & & D_n \\ | & | & & | \\ u_0 & u_1 & \cdots & u_n \\ | & | & & | \end{bmatrix}$ is a nonzero integer. Therefore C's vol-

ume is at least $(n! \prod_{l=0}^{n} |D_l|)^{-1}$. Because $|D_l| < 2^L$ for all l, $\prod_{l=0}^{n} |D_l|$
$< 2^{(n+1)L}$. But $n! < 2^L$ (because of the $n \lceil \log_2 n \rceil$ term) and thus C's
volume exceeds $2^{-(n+2)L}$. ∎

Recall that the Ellipsoid Algorithm's initial ellipsoid $E_0 = S(0,$
$n2^L)$. For all $r \geq 0$, $S(0, r) \subseteq \{ x \mid |x_j| \leq r \; \forall j \}$ and therefore the
volume of $S(0, r)$ is at most $(2r)^n$.

Theorem 48. The Ellipsoid Algorithm correctly solves LSI.

Proof. If the Ellipsoid Algorithm reports that t_l is feasible, it is obviously correct. So suppose the Ellipsoid Algorithm terminates with "infeasible" yet the problem is feasible. By Lemma 47, the set $F \cap E_0$ has volume at least $2^{-(n+2)L}$ and lies entirely within E_0. By Theorem 44(b), $F \cap E_0$ continues to be a subset of each E_l, $l = 0, 1, \ldots, K = 16n(n+1)L$. But by Theorem 44(c),

$$
\begin{aligned}
\text{volume}(E_K) &< (\text{volume}(E_0))2^{-\frac{K}{2(n+1)}} \\
&< (\text{volume}(E_0))2^{-8nL}.
\end{aligned}
$$

Volume$(E_0) < (2n2^L)^n$ because $E_0 \subseteq \left\{ x \mid |x_j| \le n2^L \right\}$. Consequently

$$
\text{volume}(E_K) < (\underbrace{2n\,2^L}_{<2^L})(2^{-8nL}) < 2^{2nL}2^{-8nL} = 2^{-6nL} < 2^{-(n+2)L},
$$

a contradiction. ∎

Throughout, we have been using exact arithmetic. In order to formally prove that linear programming is in **P**, we would have to show that a variant of the Ellipsoid Algorithm can be implemented so as to run in polynomial time on the traditional theoretical model of a computer that manipulates only integers (see chapter one). A gory and not very enlightening proof shows that this indeed can be done. One proves that each number constructed during the computation can be approximated by a rational whose binary representation is a string of $2p$ bits, p before and p after the binary point, where p is bounded above by a polynomial in L.

■ Ellipsoid Algorithm

4.5 Notes

The discovery in 1979 of the Ellipsoid Algorithm opened up the possibility that non-combinatorial methods might beat combinatorial ones for linear programming. Decades of work on Simplex had failed to yield a polynomial-time variant; in hindsight it seems that myopically jumping from vertex to neighboring vertex was the wrong strategy, at least in theory, despite its success in practice.

Unfortunately, the first proof of the polynomiality of the Ellipsoid Algorithm appeared in a four page article in the Russian journal *Doklady Akademiia Nauk USSR* [35]. As if being in Russian weren't bad enough, the proofs were omitted. The short paper in English by Gács and Lovász [19] gave the English-speaking world most of the details. This chapter is based on [19] and also on the coverage of the Ellipsoid Algorithm in Papadimitriou and Steiglitz's text [44], which was itself influenced by Aspvall and Stone [7]. Khachiyan based his algorithm on the so-called ellipsoid method of Shor [50] and Judin and Nemirovskii [33], which was developed for convex, not necessarily differentiable, programming. Khachiyan's contribution was to show that this method, when applied to linear programming, gives a polynomial-time algorithm.

The popular press couldn't help overstating the practical importance of the Ellipsoid Algorithm when it first appeared. Lawler's wonderful article "The Mathematical Sputnik of 1979" [40] is an amusing and informative account of the press coverage of the Ellipsoid Algorithm. Fortunately, coverage of Karmarkar's Algorithm (to be discussed in chapter 5) lacked most of the sensationalism that accompanied the appearance of the Ellipsoid Algorithm.

Today, the Ellipsoid Algorithm is very impractical. The Ellipsoid Algorithm is far faster than Simplex in the worst case, but this is meager consolation to those who solve linear programs in practice. A survey of results on the Ellipsoid Algorithm appears in [12].

The number of bits of precision required by the Ellipsoid Algorithm *is* bounded by a polynomial in the input size, but it's quite large. Partly for this reason, the successors to the Ellipsoid Algorithm's (e.g., the algorithms by Karmarkar [34], Renegar [45], Monteiro and Adler [42], Vaidya [60], and Gonzaga [26]) are much faster.

No one has yet found a linear programming algorithm which performs at most $q(m, n)$ arithmetic operations on operands having a polynomial number of bits, where q is some two-variable polynomial and m and n are the number of rows and columns, respectively, of the constraint matrix. Progress in the search for such a *strongly-polynomial* algorithm was made by Tardos [57]. She showed how to solve linear programs with arbitrary objective functions and right-hand sides with $q(m, n)$ arithmetic operations, provided that all the entries of the constraint matrix are bounded in absolute value (e.g., all entries are 0, 1 or -1).

Two recent books present the Ellipsoid Algorithm in full detail. *Geometric Algorithms and Combinatorial Optimization* by M. Grötschel, L. Lovász, and A. Schrijver [27] focuses on the applications of the Ellipsoid Algorithm to combinatorial optimization. Schrijver's book [47] is a voluminous reference containing everything from Simplex to Karmarkar's Algorithm, as well as numerous results on integer linear programming. Both books prove that the Ellipsoid Algorithm will solve LSI instances (and hence LP instances) if, for a vector t, we can

- determine if t is feasible, and

- if not, find a violated inequality.

Thus some linear programs with n variables and *exponentially* many constraints can be solved in time *polynomial* in n, if there exists a (possibly tricky) way to test a vector for feasibility and find a violated inequality if not. Of course one must define the linear program implicitly, since even to write it down would take exponential time.

Chapter 5

Karmarkar's Algorithm

The appearance in 1984 of Karmarkar's Algorithm for linear programming generated much excitement in the mathematical community. Also known as the *projective transformation method*, Karmarkar's Algorithm was the first polynomial-time linear programming algorithm to compete viably with Simplex on real-world problems. Like the Ellipsoid Algorithm, Karmarkar's Algorithm almost completely ignores the combinatorial structure of linear programming.

Recall that the polynomial-time Ellipsoid Algorithm generates a sequence of points *outside* the feasible region of the linear program; the Ellipsoid Algorithm is an *exterior-point method*. Simplex generates vertices of the feasible region. Unlike both, Karmarkar's Algorithm is an *interior-point method*. It generates a sequence of points inside the feasible region whose costs approach the optimal cost. At the end, it jumps to a vertex of no greater cost, which is then optimal.

5.1   Ideas

(a) Karmarkar Standard Form

Karmarkar's Algorithm directly solves only certain standard-form linear programs, those phrased in Karmarkar Standard Form. These are standard-form linear programs every constraint of which has a 0 right-hand side, except for the one constraint $x_1+x_2+\cdots+x_n = 1$. In addition, we are guaranteed that the point $a_0 = \begin{bmatrix} \dfrac{1}{n} & \dfrac{1}{n} & \cdots & \dfrac{1}{n} \end{bmatrix}^T = \dfrac{e}{n}$ is feasible, and that *every* feasible point has nonnegative cost. Our goal is simply to find a point of cost 0, or determine that none exists.

Definition.   Let $\Delta = \{\, x \in \mathbb{R}^n \mid x \geq 0,\ \sum x_j = 1 \,\}$, the $(n-1)$-*dimensional standard simplex*. Its *center* $a_0 = \dfrac{1}{n} e$.

The input to a Karmarkar Standard Form linear program consists of an $m \times n$ matrix A of integers and a vector $c \in \mathbb{Z}^n$. Where $\Omega = \{x \in \mathbb{R}^n \mid Ax = 0\}$, the feasible region is $\Pi = \Omega \cap \Delta$. The cost $c^T x$ of every $x \in \Pi$ is nonnegative, and $a_0 \in \Pi$. The goal is to find a point in Π of cost 0. If none exists, we need not find a minimum-cost point in Π.

Without loss of generality, we will assume that all $m+1$ equality constraints are linearly independent, since we can eliminate redundant rows from A in order to make this true.

Karmarkar Standard Form linear programs are special cases of arbitrary standard-form linear programs. However, after we finish describing and analyzing Karmarkar's Algorithm, we will go back and show how, in polynomial time, any linear program can be converted to a Karmarkar Standard Form linear program. Thus Karmarkar Standard Form linear programs are completely general.

(b) "Roundness" Bounds

The intersection of an affine space of nonzero dimension and a sphere centered in the affine space is a "lower-dimensional sphere" of the same radius. For example, the intersection of a plane in $\mathbb{R}^3$ and a 3-dimensional sphere centered in the plane is a disk, a 2-dimensional "sphere." Suppose our goal were to minimize $c^T x$ over $S' = \{x \mid Ax = b\} \cap S(v, r)$ (where $Av = b$), the intersection of an affine space and a sphere centered in the space. This is easy: Let c_P be the projection of c onto $\{x \mid Ax = 0\}$; $-c_P$ is the direction "within" the affine space movement along which decreases the objective function most rapidly. Start at the center v and take a step of length r in direction $-c_P$, just reaching a point on the surface of $S(v, r)$. That point is of minimum cost. (We will prove this shortly.)

Example. To minimize $3x_1 + 4x_2 = [3 \ \ 4]^T \begin{bmatrix} x_1 \\ x_2 \end{bmatrix}$ over $\{x \in \mathbb{R}^2 \mid (x_1 - 2)^2 + (x_2 - 5)^2 \le 16\}$—here the affine space is all of $\mathbb{R}^2$—we start at the center $\begin{bmatrix} 2 \\ 5 \end{bmatrix}$ and take a step of length 4 in the direction $\begin{bmatrix} -3 \\ -4 \end{bmatrix}$:

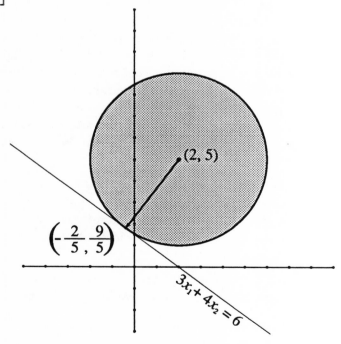

If linear programming were the problem of minimizing a linear function over the intersection of an affine space and a sphere centered in the affine space, it would be an easy problem. How can we use this observation to help us solve linear programs? Suppose we have found a feasible point z. Since it is so easy to minimize $c^T x$ over a "lower-dimensional sphere" which is the intersection of an affine space and a sphere centered in it, why don't we embed inside the feasible region a "lower-dimensional sphere" B centered at z, and minimize $c^T x$ over B. We would want B to be as large as possible and still fit inside the feasible region. If B isn't much smaller than the feasible region itself, the minimum over B ought not be much larger than the true minimum. If w is a minimum-cost point in B, then the cost of w should be significantly smaller than the cost of z. We could then repeat this process, starting from w.

How good an approximate answer would we get? Intuitively, if the feasible region is "round," like a sphere, with its "center" at z, so that the inscribed ball includes most of the polytope, then the minimum over the inscribed ball should be a good approximation, whereas if it's long and narrow, like a triangle with side lengths 1, 1000 and 1000, the inscribed ball would be too small and the answer would be worthless. A natural measure of "roundness" is the ratio of the radius of the smallest sphere which contains the feasible region to the radius of the largest sphere inside it. For a sphere we get a ratio of one, the best possible ratio. Long and narrow isosceles triangles give very large ratios, suggesting that the approximation will be almost useless.

Now let us formalize these arguments. Let P be any polytope and $z \in P$. Let $E \subseteq P$ be any closed convex body also containing z. (Think of E as an inscribed sphere centered at z.) If we minimize a linear function $h(x) = c^T x$ over E instead of over the larger, complicated set P, how good a solution do we get?

To derive a bound, "scale up" the body E about z by a sufficiently large factor ν (if possible) so that the new body E' contains P and is of the same shape as E:

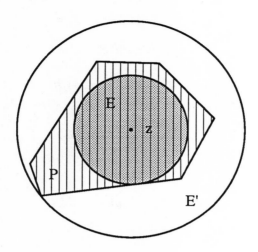

E' has the same shape as E but is ν times as large: $E' = \{z+\nu(x-z) \mid x \in E\}$. $E \subseteq P \subseteq E'$.

If $x \in E$, let $x' = z + \nu(x - z)$, so that $x' \in E'$. Define f by $f(x) = h(z) - h(x)$. By linearity of h,

$$
\begin{aligned}
h(x') - h(z) &= \nu h(x - z) \\
&= \nu[h(x) - h(z)].
\end{aligned}
$$

Thus

$$
\begin{aligned}
f(x') &= h(z) - h(x') \\
&= \nu[h(z) - h(x)] \\
&= \nu f(x),
\end{aligned}
$$

so that for every $x \in E$, there is an $x' \in E'$ so that $f(x') = \nu f(x)$. Similarly, one can argue that for every $x' \in E'$, there is an $x \in E$ satisfying $f(x') = \nu f(x)$. Thus the maximum value of f over E' is exactly ν times the maximum value of f over E. Now let h_E and $h_{E'}$ denote the minimum values of h over E and E', respectively. Since

$h(z) - h_{E'}$ is the maximum value of f over E' and $h(z) - h_E$ is the maximum value of f over E, we conclude that

$$h(z) - h_{E'} = \nu[h(z) - h_E].$$

Of course the number we really want is h_P, the minimum of h over polytope P. Because $E \subseteq P \subseteq E'$, $h_E \geq h_P \geq h_{E'}$. Hence

$$\begin{aligned} h(z) - h_P &\leq& h(z) - h_{E'} \\ &=& \nu[h(z) - h_E]. \end{aligned}$$

Therefore

$$(\nu - 1)h(z) - \nu h_E \geq -h_P$$

and

$$\begin{aligned} (\nu - 1)h(z) - (\nu - 1)h_P &\geq& \nu h_E - \nu h_P; \\ (\nu - 1)[h(z) - h_P] &\geq& \nu(h_E - h_P). \end{aligned}$$

If $h(z) > h_P$,

$$\frac{h_E - h_P}{h(z) - h_P} \leq 1 - \frac{1}{\nu}.$$

If $w \in E$ satisfies $h(w) = h_E$,

$$\frac{h(w) - h_P}{h(z) - h_P} \leq 1 - \frac{1}{\nu}.$$

The difference $h(z) - h_P$ is a measure of how "bad" z is, how much larger its cost is than the optimal cost, h_P. Point w is better than z in that its "excess cost" is only $1 - 1/\nu$ times as great as z's. The smaller ν is, the more progress we make in switching from z to w. The more closely P resembles a sphere centered at z, the smaller ν is: we want "rounded" polytopes. In fact, Karmarkar chose the feasible region to be the intersection of an affine space and a simplex specifically because (as we shall soon see) a simplex is a rounded polytope!

Karmarkar's Algorithm starts at $z = a_0$ and generates a sequence of feasible points of cost approaching 0 (provided that the optimal cost is 0). Suppose the current point is $a \in \Pi$ and $c^T a > 0$. How shall we generate the next point, b? We might inscribe in Π a ball B centered at a, and minimize $c^T x$ over B. But there is no reason to expect the minimum of $c^T x$ over B even to remotely resemble the true minimum of $c^T x$ over Π, since Π is probably not rounded about a. For example, a might be very close to the boundary of Π. Karmarkar's brilliant solution to this dilemma—this is the key to the whole algorithm—is to apply a transformation T_a to the problem to get a new problem in which a is replaced by a_0. The feasible region Π' of the new problem, like Π, will be the intersection of Δ and an affine space, and Δ *is rounded about* a_0. This means that by minimizing in the new problem over a ball inscribed in Π', we will get a point b' whose cost in the new problem is significantly smaller than that of a_0. It is then a simple matter to apply to b' the inverse of T_a, getting a point b. Because the cost of b' in the new problem is much less than that of a_0, in the original problem the cost of b will be much less than that of a. Repeated application of this procedure will give us a sequence of points whose costs approach 0. (Actually, the situation is a bit more complicated than this description.)

To what extent is the $(n-1)$-dimensional standard simplex Δ "rounded?"

Definition. Let $B(a_0, r)$ be the largest $(n-1)$-dimensional ball centered at a_0 in Δ, and let $B(a_0, R)$ be the smallest $(n-1)$-dimensional ball centered at a_0 and containing Δ.

A diagram of $B(a_0, r)$, Δ, and $B(a_0, R)$ appears after the proof of Lemma 49.

Lemma 49. The radii $r = \dfrac{1}{\sqrt{n(n-1)}}$ and $R = \sqrt{\dfrac{n-1}{n}}$.

Proof.

Inscribed Ball. We need this claim.

Claim. Suppose $y \in \mathbb{R}^n$ satisfies $y_l \geq 1/n$ for $l < n$, $y_n = 0$ and $y_1 + y_2 + \cdots + y_n \geq 1$. Then $\|y - a_0\|^2 \geq r^2 = \dfrac{1}{n(n-1)}$.

Proof of Claim. Let $z_l = y_l - 1/n$ for $l = 1, 2, ..., n-1$.

$$\|y - a_0\|^2 = \sum_{l=1}^{n-1} \left(y_l - \frac{1}{n} \right)^2 + \frac{1}{n^2}$$

$$= \sum_{l=1}^{n-1} z_l^2 + \frac{1}{n^2}.$$

All $z_l \geq 0$ and $\sum_{l=1}^{n-1} z_l \geq 1 - (n-1)/n = 1/n$. Subject to the constraint $\sum_{l=1}^{n-1} z_l \geq 1/n$, $\sum_{l=1}^{n-1} z_l^2 + 1/n^2$ is minimized when $\sum_{l=1}^{n-1} z_l = 1/n$ and all the z_l's are identical, in which case

$$\sum_{l=1}^{n-1} z_l^2 + \frac{1}{n^2} = \frac{1}{n^2(n-1)} + \frac{1}{n^2} = \frac{1}{n(n-1)} = r^2.$$

This completes the proof of the claim.

The distance from a_0 to the point $\left[0 \quad \dfrac{1}{n-1} \quad \dfrac{1}{n-1} \quad \cdots \quad \dfrac{1}{n-1} \right]^T$ on the boundary of Δ is

$$\sqrt{\frac{1}{n^2} + (n-1) \left(\frac{1}{n-1} - \frac{1}{n} \right)^2} = \sqrt{\frac{1}{n^2} + \frac{n-1}{(n-1)^2 n^2}} = r,$$

so the radius of any inscribed $(n-1)$-dimensional ball cannot exceed r.

Upper bound on r in hand, let us now prove that every point x satisfying $e^T x = 1$ which is at distance less than r from a_0 is actually in Δ. (Those exactly r from a_0 will then lie in Δ as well.) For a contradiction, suppose that $e^T x = 1$ and $\|x - a_0\| < r$, yet some entry, say, the last, is negative. Define $y \in \mathbb{R}^n$ as follows.

$$y_l = \begin{cases} x_l & \text{if } l < n \text{ and } x_l \geq \dfrac{1}{n} \\[2mm] \dfrac{2}{n} - x_l & \text{if } l < n \text{ and } x_l < \dfrac{1}{n} \\[2mm] 0 & \text{if } l = n \end{cases}.$$

It is easy to verify that $y_l \geq 1/n$ for $l < n$, $y_n = 0$, $e^T y \geq 1$ and $\|y - a_0\| = \|x - a_0\| < r$. This contradicts the claim, so we are done.

Circumscribed Ball. The distance from a_0 to any one of Δ's vertices, say, the vertex $[1 \ 0 \ 0 \ \cdots \ 0]^T$, is $\sqrt{(1 - 1/n)^2 + (n-1)/n^2} = \sqrt{(n-1)/n} = R$. Thus every circumscribing $(n-1)$-dimensional ball has radius at least R. To complete the proof, we show that the distance from any point in Δ to a_0 is at most R.

If α and β are nonnegative reals, it is easy to verify that $(\alpha - 1/n)^2 + (\beta - 1/n)^2 \leq ((\alpha + \beta) - 1/n)^2 + (0 - 1/n)^2$. Let $x \in \Delta$. For $l = 2, 3, ..., n$, iteratively replace entries x_1 and x_l of x by $x_1 + x_l$ and 0, respectively. Doing so can only increase the distance from a_0 and leaves the point in Δ. Eventually, we will be left with the vertex $[1 \ 0 \ 0 \ \cdots \ 0]^T$. Since its distance from a_0 is R, no distance from any point in Δ to a_0 can exceed R. ∎

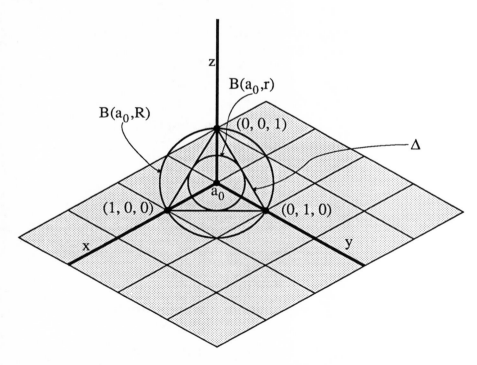

For Δ and $z = a_0$, we can take $\nu = R/r = n - 1$. The simplex is "rounded" about its center, as R/r is only $n-1$.

(c) Projective Transformations

Somehow we must construct a transformation T_a that converts the problem to a new one in which a is replaced by a_0. We want the image of every point to lie in Δ, so we need the sum of the coordinates of $T_a(x)$ to be one. Since $a = [a_1 \ a_2 \ \cdots \ a_n]^T$, an obvious transformation which maps a to a_0 is to map x to $\left[\dfrac{x_1}{a_1} \ \dfrac{x_2}{a_2} \ \cdots \ \dfrac{x_n}{a_n}\right]^T$, and then divide by the sum of the coordinates so as to make the final sum unity. (We must ensure that $a > 0$ for this mapping to make sense.) In other words, $T_a(x) = x'$ where

$$x'_j = \frac{x_j/a_j}{\sum_l x_l/a_l}.$$

These are the transformations Karmarkar uses! Equivalently,

$$T_a(x) = \frac{Cx}{e^T C x}, \text{ where } C = \begin{bmatrix} a_1^{-1} & 0 & 0 & 0 \\ 0 & a_2^{-1} & 0 & 0 \\ 0 & 0 & \ddots & 0 \\ 0 & 0 & 0 & a_n^{-1} \end{bmatrix}.$$

Let

$$D = C^{-1} = \begin{bmatrix} a_1 & 0 & 0 & 0 \\ 0 & a_2 & 0 & 0 \\ 0 & 0 & \ddots & 0 \\ 0 & 0 & 0 & a_n \end{bmatrix}$$

so that for all x, $T_a(x) = \dfrac{D^{-1}x}{e^T D^{-1} x}$.

Transformation T_a is a special kind of projective transformation.

Definition. A *projective transformation* of $\mathbb{R}^n$ maps $x \mapsto \dfrac{Ex + d}{f^T x + g}$, where E is an $n \times n$ matrix, $d, f \in \mathbb{R}^n$, $g \in \mathbb{R}$ and the $(n+1) \times (n+1)$

matrix

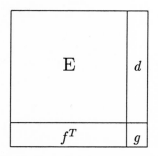

is nonsingular. The transformation is defined at x if and only if $f^T x + g \neq 0$.

Let E be an arbitrary $n \times n$ nonsingular matrix. Define $d = 0$, $g = 1$ and $f^T = e^T(E - I)$. Then, because E is nonsingular, $d = 0$ and $g \neq 0$, the map $x \mapsto \dfrac{Ex}{f^T x + 1} = \dfrac{Ex + d}{f^T x + g}$ is a projective transformation. Let Σ denote $\{x | e^T x = 1\}$. For $x \in \Sigma$,

$$
\begin{aligned}
\frac{Ex + d}{f^T x + g} &= \frac{Ex}{e^T(E - I)x + 1} \\
&= \frac{Ex}{e^T Ex - e^T x + 1} \\
&= \frac{Ex}{e^T Ex}.
\end{aligned}
$$

In other words, for $x \in \Sigma$, the mapping $x \mapsto \dfrac{Ex}{e^T Ex}$ for any nonsingular E is a restriction to Σ of a projective transformation. In our case, because $a > 0$, C is a diagonal matrix with a positive diagonal, and so is clearly nonsingular.

Transformation T_a indeeds maps the simplex Δ into Δ: If x is in Δ, then $x \geq 0$ and therefore $T_a(x) \geq 0$. Furthermore, the sum of the entries in $T_a(x)$ is $e^T T_a(x) = \dfrac{e^T C x}{e^T C x} = 1$. In fact, T_a maps Δ *onto* Δ, because the function $x' \mapsto \dfrac{C^{-1} x'}{e^T C^{-1} x'}$ is the inverse of T_a. In other words, if $T_a(x) = x'$, then $x_j = \dfrac{a_j x'_j}{\sum_l a_l x'_l}$. It follows that

$$\sum_j A_j x_j = 0 \iff \sum_j A_j \left[\frac{a_j x'_j}{\sum_l a_l x'_l} \right] = 0 \iff \sum_j (A_j a_j) x'_j = 0.$$

Because AD is the $m \times n$ matrix whose jth column is $a_j A_j$, $Ax = 0$ if and only if $ADx' = 0$. Recall that $\Omega = \{ x \mid Ax = 0 \}$. Letting $\Omega' = \{ x' \in \mathbb{R}^n \mid ADx' = 0 \}$, we see that $T_a(\Omega) = \Omega'$. Because $T_a(a) = a_0$, if $a \in \Omega$ then $a_0 \in \Omega'$.

Since $T_a(\Delta) = \Delta$ and $T_a(\Omega) = \Omega'$ and T_a is invertible, $T_a(\Delta \cap \Omega) = \Delta \cap \Omega'$. This means that the image of the original feasible region Π is $\Delta \cap \Omega'$. Define Π' to be $\Delta \cap \Omega'$ so that $T_a(\Pi) = \Pi'$.

Is Π', the image of Π, a rounded polytope? If not, all this work is for naught. Recall that $B(a_0, r)$ is the largest $(n-1)$-dimensional ball inscribed in Δ and $B(a_0, R)$ is the smallest $(n-1)$-dimensional circumscribing ball; $B(a_0, r) \subseteq \Delta \subseteq B(a_0, R)$. Therefore

$$B(a_0, r) \cap \Omega' \subseteq \Delta \cap \Omega' \subseteq B(a_0, R) \cap \Omega'.$$

Since Ω' is an affine space containing a_0, $B(a_0, r) \cap \Omega'$ and $B(a_0, R) \cap \Omega'$ are balls of dimension at most $n - 1$, the radii of which are respectively r and R. Define

$$B'(a_0, r) = B(a_0, r) \cap \Omega'$$

and

$$B'(a_0, R) = B(a_0, R) \cap \Omega'.$$

Then

$$B'(a_0, r) \subseteq \Pi' \subseteq B'(a_0, R)$$

and R/r is only $n - 1$. Therefore Π' is a rounded polytope! By the "roundness" results, to approximately minimize a linear function over Π', we could minimize over the inscribed ball $B'(a_0, r)$ instead of Π', and "lessen the gap" by a factor of $1 - \dfrac{1}{\nu} = 1 - \dfrac{1}{n-1}$.

The bad news is that there is one serious complication: the original linear objective function becomes nonlinear under this mapping. Our original goal was to minimize $c^T x$ over $x \in \Pi$. Because

$$c^T x = \frac{\sum_j (c_j a_j) x'_j}{\sum_l a_l x'_l},$$

to minimize $c^T x$ over $x \in \Pi$ we should minimize $\dfrac{\sum (c_j a_j) x'_j}{\sum a_l x'_l}$ over $x' \in \Pi'$. This is a *nonlinear* function of x'. The outline above of Karmarkar's Algorithm is simplistic, since the transformed problem is really a problem of minimizing a rational function over a polytope. In the next section we shall see that it suffices to approximately minimize the linear numerator $\sum_j (c_j a_j) x'_j$ of the transformed objective function.

(d) A Nearly Invariant Potential Function

Karmarkar's solution to this dilemma is to argue that by approximately minimizing $\sum_j (c_j a_j) x'_j$, we will still eventually reach a global minimum. Though the costs of the points generated will approach 0, they will not decrease monotonically. (The outline above is incorrect in suggesting that the costs decrease monotonically.) We will prove, however, by exhibiting a *potential function*, that over a long sequence of iterations, the cost will decrease dramatically, though in any single step it might increase.

Where $c^T x$ is our objective function, $x \in \Pi$, $c^T x > 0$ and $x > 0$, define $f(x) = \ln \left[\dfrac{(c^T x)^n}{x_1 \cdots x_n} \right]$, a *potential function* which quantifies how much progress we have made. In particular, $(c^T x)^n \leq e^{f(x)}$ and therefore $c^T x \leq e^{f(x)/n}$. If $f(x)$ approaches $-\infty$, $c^T x$ must approach 0. The potential function f will be used to measure our progress—if $f(x)$ is very small, $c^T x$ is very close to 0. Also, we will show that to drive $f(x)$ down, it suffices to approximately minimize a linear function. We will see that always producing a new point b for which $f(b) \leq f(a) - \delta$ (for a constant $\delta > 0$) is enough to guarantee polynomial-time convergence.

In order to compute f, we will augment our RAM's instruction set with the natural logarithm, a function not included among a RAM's basic instruction set. Numerical subroutines exist which approximate the natural logarithm very well. Or, instead of approximating the logarithm, we could easily rewrite Karmarkar's Algorithm—though we won't—so as to eliminate all logarithms. Only rational operations and square roots would then be needed.

Recall that

$$D = C^{-1} = \begin{bmatrix} a_1 & 0 & 0 & 0 \\ 0 & a_2 & 0 & 0 \\ 0 & 0 & \ddots & 0 \\ 0 & 0 & 0 & a_n \end{bmatrix}$$

so that

$$T_a(x) = \frac{D^{-1}x}{e^T D^{-1}x}.$$

Let

$$c' = Dc = \begin{bmatrix} a_1 c_1 \\ a_2 c_2 \\ \vdots \\ a_n c_n \end{bmatrix}.$$

Now define

$$f'(y) = \ln\left[\frac{(c'^T y)^n}{y_1 \cdots y_n}\right]$$

for $y \in \Pi'$, $c'^T y > 0$, $y > 0$. Recall that to minimize $c^T x$ over Π, we should minimize $\dfrac{\sum(c_j a_j)y_j}{\sum a_l y_l}$ over $y \in \Pi'$. Hence f' is what the potential function would be in the transformed problem, if its objective function were simply $c'^T y$.

The next lemma says that the potential function is almost invariant under T_a: $f'(T_a(x)) - f(x)$ is independent of x.

Lemma 50.   For all $x \in \Pi$, if $c^T x > 0$ and $x > 0$, then

$$f'(T_a(x)) = f(x) + \ln(a_1 a_2 \cdots a_n).$$

Proof. If $c^T x > 0$ and $x > 0$, then both $f(x)$ and $f'(T_a(x))$ exist. Because

$$T_a(x) = \frac{D^{-1}x}{e^T D^{-1}x}$$

and

$$f'(y) = \ln \frac{(c'^T y)^n}{y_1 \cdots y_n},$$

$$
f'(T_a(x)) = \ln \frac{\left[c^T D^T \left(\dfrac{D^{-1}x}{e^T D^{-1}x} \right) \right]^n}{\dfrac{\dfrac{x_1}{a_1} \dfrac{x_2}{a_2} \cdots \dfrac{x_n}{a_n}}{(e^T D^{-1}x)^n}} - \ln(a_1 a_2 \cdots a_n)
$$

$$= f(x). \quad \blacksquare$$

Our Strategy. Given $a \in \Pi$, first use T_a to projectively map a to the center a_0. Using the roundness theorems, (somehow) find a point $b' \in \Pi'$ for which $f'(b') \leq f'(a_0) - \delta$. Then apply the inverse of T_a to b', getting $b \in \Pi$. The "almost invariance" of f implies that $f(b) \leq f(a) - \delta$:

$$
\begin{aligned}
f(b) &= f'(b') - \ln(a_1 \cdots a_n) \\
&\leq (f'(a_0) - \delta) - \ln(a_1 \cdots a_n) \\
&= f(a) - \delta.
\end{aligned}
$$

This is progress! We will drive f toward $-\infty$ by making many iterations, each of which decreases f by δ.

5.2 The Algorithm

As we did for the Ellipsoid Algorithm, let $L = (m+1)n + \lceil \log_2 |P| \rceil + n \lceil \log_2 n \rceil$, where P is the product of the nonzero entries in c and among all $m + 1$ constraints.

Notation. Let $\mathcal{P}$ denote the Karmarkar Standard Form linear program we wish to solve.

Let α be any real strictly between 0 and 1/2, and let

$$\delta = \alpha - \frac{\alpha^2}{1-\alpha} > 0.$$

Karmarkar's Algorithm will run in polynomial-time for any $\alpha \in (0, 1/2)$.

Here is the main body of Karmarkar's Algorithm. To generate the next point in the sequence, it calls function Φ whose code will be given shortly. Φ takes as input a positive point $a \in \Pi$ of positive cost and generates a positive point $\Phi(a) \in \Pi$; provided that the optimal cost is 0, either the cost of $\Phi(a)$ is 0 or $f(\Phi(a)) \leq f(a) - \delta$.

Karmarkar's Algorithm

1. Let $x^{(0)} = a_0 = \frac{1}{n}e$ and let $K = \lceil 2nL/\delta \rceil$ (the iteration bound).
   If $c^T a_0 = 0$, return a_0.
2. For $k = 1$ to K do
   Use Function Φ to generate $x^{(k)} = \Phi(x^{(k-1)})$.
   If $c^T x^{(k)} = 0$, return $x^{(k)}$ and halt.
   If $f(x^{(k)}) > f(x^{(k-1)}) - \delta$, report that the optimal cost is
   positive and halt.
3. Find a vertex v of cost no more than $c^T x^{(K)}$.
   Return v, which must have cost 0.

End.

Theorem 51. Suppose that Φ always returns a positive point $\Phi(a)$ in Π if $a \in \Pi$ is positive and of positive cost, and that if the minimum cost in $\mathcal{P}$ is 0, then either the cost of $\Phi(a)$ is 0 or $f(\Phi(a)) \leq f(a) - \delta$. Then Karmarkar's Algorithm correctly solves Karmarkar Standard Form linear programs.

Proof. The only way Karmarkar's Algorithm can report that the optimal cost is positive is if for some k, the cost of $x^{(k)}$ is positive and $f(x^{(k)}) > f(x^{(k-1)}) - \delta$, and in this case we know from the hypothesis on Φ that the optimal cost is positive.

If Karmarkar's Algorithm returns $x^{(k)}$ after verifying that $c^T x^{(k)} = 0$, clearly it is correct.

So suppose that $f(x^{(k)}) \leq f(x^{(k-1)}) - \delta$ for $k = 1, 2, ..., K$. Then

$$f(x^{(K)}) \leq f(x^{(0)}) - \left(\frac{2nL}{\delta}\right)\delta = f(x^{(0)}) - 2nL$$

and

$$c^T x^{(K)} \leq e^{f(x^{(K)})/n} \leq e^{[f(x^{(0)}) - 2nL]/n}.$$

From $x^{(0)} = a_0$ we infer that

$$
\begin{aligned}
f(x^{(0)}) &= \ln\left[\frac{(c^T a_0)^n}{(1/n)^n}\right] \\
&= n\ln(nc^T a_0) \\
&= n\ln(c_1 + c_2 + \cdots + c_n).
\end{aligned}
$$

But $\ln(c_1 + c_2 + \cdots + c_n) \leq L$ (exercise), so $f(x^{(0)}) \leq nL$. Consequently $\dfrac{f(x^{(0)}) - 2nL}{n} \leq -L$. Therefore $c^T x^{(K)} \leq e^{-L} < 2^{-L}$. Let v be the vertex returned by Karmarkar's Algorithm in this case. Lemma 33 tells us that vertex v is a rational vector with a positive common denominator less than 2^L; *ergo* $|c^T v| > 2^{-L}$ if $c^T v > 0$. Since $0 \leq c^T v \leq c^T x^{(K)} < 2^{-L}$, we must have $c^T v = 0$. Therefore Karmarkar's Algorithm returns a point of cost 0 when one exists, and otherwise reports the nonexistence of such a point. ∎

We must exhibit Φ and prove that either $c^T \Phi(a) = 0$ or $f(\Phi(a)) \leq f(a) - \delta$ (if the optimal cost is 0).

Crucial Idea. Given $a \in \Pi$, our goal is to projectively map a to a_0, find a $b' \in \Pi'$ such that $f'(b') \leq f'(a_0) - \delta$, and then let b be the inverse of b'. How can we construct b'? Recall that

$$f'(y) = \ln \frac{(c'^T y)^n}{y_1 y_2 \cdots y_n}.$$

Since nonlinear functions such as f' are much more difficult to deal with than linear functions such as $c'^T y$, and since Π' is rounded about a_0, it seems natural to let b' be the point that minimizes $c'^T y$ over a ball inscribed in Π'. In other words, to minimize the numerator of $e^{f'(y)}$ over an inscribed ball and hope that the denominator doesn't inadvertently shrink too much. If we were to minimize over the largest ball which can be inscribed in Π' (this is a ball of radius r), the denominator indeed might shrink too much. One or more of the y_j's could become so small that the net effect would be an increase in f'. Instead, we minimize $c'^T y$ over a ball of radius α/n inscribed in Π'. Since $r = \dfrac{1}{\sqrt{n(n-1)}}$, α/n is just slightly less than αr. That is, we let b' be a point which minimizes $c'^T y$ over $S(a_0, \alpha/n) \cap \Pi'$. With this choice of b', we will be able to show that $f'(b') \le f'(a_0) - \delta$.

Now we show that $S(a_0, \alpha/n) \cap \Pi'$ is just the disguised intersection of the sphere $S(a_0, \alpha/n)$ and the affine space $\Sigma \cap \Omega'$ (where Σ denotes $\{x | e^T x = 1\}$); since we already know how to minimize a linear function over the intersection of an affine space and a sphere centered in it, finding b' will be easy. Because every point in $S(a_0, \alpha/n)$ has positive coordinates,

$$S(a_0, \alpha/n) \cap \Delta = S(a_0, \alpha/n) \cap \Sigma.$$

This means that

$$
\begin{aligned}
S(a_0, \alpha/n) \cap \Pi' &= S(a_0, \alpha/n) \cap \underbrace{[\Delta \cap \Omega']}_{\Pi'} \\
&= S(a_0, \alpha/n) \cap [\Sigma \cap \Omega'].
\end{aligned}
$$

Notation. Let $\overline{\Omega} = \Omega' \cap \Sigma$, an affine space containing a_0. Let B be the $(m+1) \times n$ matrix obtained by adding a row of ones below AD:
$$B = \begin{bmatrix} AD \\ e^T \end{bmatrix}.$$

Since $\Omega' = \{x | ADx = 0\}$, $\overline{\Omega} = \{x | Bx = e'\}$, where e' is the $(m+1)$-vector which is all 0 except for a one in the last position.

Now we are ready to present Φ. Function Φ does nothing more than (1) transform the problem, taking a to $a_0 = T_a(a)$, (2) find a point b' which minimizes $c'^T y$ over $S(a_0, \alpha/n) \cap \overline{\Omega}$, and (3) return the inverse image b of b'. Either $c^T b = 0$ or $f(b) \leq f(a) - \delta$ (provided that the minimum cost in $\mathcal{P}$ is 0).

Function $\Phi(a)$

Where $a \in \Pi$, $a > 0$ and $c^T a > 0$, the output vector $b = \Phi(a)$ is in Π and $b > 0$. If the minimum cost in $\mathcal{P}$ is 0, then either $c^T b = 0$ or $f(b) \leq f(a) - \delta$.

1. Let

$$D = \begin{bmatrix} a_1 & 0 & 0 & 0 \\ 0 & a_2 & 0 & 0 \\ 0 & 0 & \ddots & 0 \\ 0 & 0 & 0 & a_n \end{bmatrix}.$$

For all x, $T_a(x) = \dfrac{D^{-1}x}{e^T D^{-1}x}$ and $T_a^{-1}(x') = \dfrac{Dx'}{e^T Dx'}$.

2. Let $c' = Dc$.

3. Construct B by adding a row of 1's below AD: $B = \begin{bmatrix} AD \\ e^T \end{bmatrix}$.

   To find $b' \in \Pi'$ such that $f'(b') \leq f'(a_0) - \delta$, we will minimize $c'^T y$ over $y \in S(a_0, \alpha/n) \cap \overline{\Omega}$, an easy task since $\overline{\Omega}$ is affine and $a_0 \in \overline{\Omega}$.

4. Let c_P be the projection of c' onto B's nullspace:

$$c_P = c' - B^T (BB^T)^{-1} Bc'.$$

   We prove in Lemma 52 that BB^T is invertible and that c_P is the projection of c' onto the nullspace of B.

5. If $c_P = 0$, let $\hat{c}_P = 0$. Otherwise, let $\hat{c}_P = \dfrac{c_P}{\|c_P\|}$, a unit vector in the same direction as c_P.

6. Take a step of length α/n in direction $-c_P$ to get b':

$$b' = a_0 - \frac{\alpha}{n}\hat{c}_P.$$

Lemma 53 proves that b' minimizes $c'^T y$ over $y \in S(a_0, \alpha/n) \cap \overline{\Omega}$.

7. Now apply T_a^{-1} to b' to get b:

Return

$$b = \frac{Db'}{e^T Db'}.$$

End.

5.3   Analysis

Lemma 52.   Step 4 of Karmarkar's Algorithm correctly computes the projection of c' onto the nullspace of B.

Proof. D being invertible, the rank of the $m \times n$ matrix AD is exactly the same as the rank of A, which is m. The last row e^T of B cannot be dependent on the other m, for otherwise $(AD)a_0 = 0$ would imply that $e^T a_0 = 0$, a contradiction. Therefore the rank of B is $m + 1$.

We must show that the rank of the $(m + 1) \times (m + 1)$ matrix BB^T is $m+1$, so that $(BB^T)^{-1}$ exists. For a contradiction, suppose that y is a nonzero $(m + 1)$-vector for which $(BB^T)y = 0$. Then $y^T(BB^T)y = 0$. But this is $(B^T y)^T(B^T y) = \|B^T y\|^2$ and hence $B^T y = 0$. This is a nontrivial linear combination of the rows of B, violating the fact that B has full row rank.

Clearly c_P is in the nullspace of B. A product of the form B^T times a vector, $c' - c_P$ is in the row space of B. Since c' has been written as the sum of a vector in the nullspace of B and a vector in the rowspace of B, the former must be the projection onto the nullspace.   ∎

Lemma 53 validates our method of minimizing a linear function over the intersection of a ball and an affine space.

Lemma 53. The point b' obtained by Φ minimizes $c'^T x$ over $S(a_0, \alpha/n) \cap \overline{\Omega}$.

Proof. The point b' is in $\overline{\Omega}$. Let $z \in S(a_0, \alpha/n) \cap \overline{\Omega}$. Let e' be the $(m+1)$-vector which is all 0, except for a 1 in the last position. Because $b', z \in \overline{\Omega}$, it follows that $Bb' = e'$ and $Bz = e'$, and therefore $B(b' - z) = 0$. Since c_P is the projection of c' onto the nullspace of B, $c' - c_P$ is orthogonal to every vector in the nullspace of B, e.g., $b' - z$. Thus $(c' - c_P)^T(b' - z) = 0$, which implies that

$$c'^T(b' - z) = c_P^T(b' - z).$$

$$c_P^T(b' - z) = \|c_P^T\| \cdot \hat{c}_P^T [\underbrace{a_0 - \frac{\alpha}{n}\hat{c}_P}_{b'} - z]$$

$$= \|c_P^T\| \left[\hat{c}_P^T(a_0 - z) - \frac{\alpha}{n} \right]$$

The Cauchy-Schwarz inequality tells us that $\hat{c}_P^T(a_0 - z) \leq \|\hat{c}_P\| \cdot \|a_0 - z\| \leq \|a_0 - z\|$ (because $\|\hat{c}_P\|$ is 0 or 1), which is at most α/n because $z \in S(a_0, \alpha/n)$. Hence $c_P^T(b' - z) \leq 0$, and therefore $c'^T(b' - z) \leq 0$. It follows that $c'^T b' \leq c'^T z$ for all $z \in S(a_0, \alpha/n) \cap \overline{\Omega}$. ∎

Lemma 54. Let b' be any point which minimizes $c'^T y$ over $\overline{\Omega} \cap S(a_0, \alpha/n) = \Pi' \cap S(a_0, \alpha/n)$. Assume that the minimum cost in $\mathcal{P}$ is 0, and that $c'^T a_0 > 0$. Then

$$\frac{c'^T b'}{c'^T a_0} \leq 1 - \frac{\alpha/n}{R}$$

and

$$\left(\frac{c'^T b'}{c'^T a_0} \right)^n \leq e^{-\alpha/R}.$$

$(R = \sqrt{(n-1)/n}$ is the radius of the smallest ball that can be circumscribed about Δ.)

Proof.   Because $\overline{\Omega}$ is an affine space, $S(a_0, \alpha/n) \cap \overline{\Omega} \subseteq \Pi' \subseteq S(a_0, R) \cap \overline{\Omega}$, and $a_0 \in \overline{\Omega}$, we can apply the roundness bounds with $E = S(a_0, \alpha/n) \cap \overline{\Omega}$, $P = \Pi'$ and $E' = S(a_0, R) \cap \overline{\Omega}$. The scale factor $\nu = \dfrac{R}{\alpha/n}$. By hypothesis the minimum value of $c'^T y$ over $y \in \Pi'$ is 0 (because the minimum cost in $\mathcal{P}$ is 0). We infer from the roundness bounds that

$$\frac{c'^T b' - 0}{c'^T a_0 - 0} \le 1 - \frac{1}{\nu} = 1 - \frac{\alpha/n}{R}.$$

From the fact that $1 + t \le e^t$ for all real t, we infer that

$$\frac{c'^T b'}{c'^T a_0} \le 1 - \frac{\alpha/n}{R} \le e^{-\alpha/(Rn)}$$

and therefore

$$\left(\frac{c'^T b'}{c'^T a_0} \right)^n \le e^{-\alpha/R}. \quad \blacksquare$$

Now we show that $f(b)$ is significantly smaller than $f(a)$.

Theorem 55.   Let b be the point obtained in the last line of Φ. Assume that $w \in \Pi$ satisfies $c^T w = 0$ (w is unknown, of course). Then either $c^T b = 0$ or $f(b) \le f(a) - \delta$.

Proof.   We first argue that under the hypothesis that $c^T w = 0$, c_P cannot be 0. For suppose $c_P = 0$. Let $y \in \Pi'$ be arbitrary. Because $c_P = 0$, c' is in the rowspace of B and is orthogonal to every vector in the nullspace of B, in particular to $y - a_0$. Thus $c'^T y = c'^T a_0$. Now $c'^T a_0 > 0$, for otherwise $c'^T a = 0$ and the algorithm would have halted before reaching this point. Because $c^T w = 0$, it follows that $c'^T w' = 0$ (where $w' = T_a(w)$). But since $c'^T w' = c'^T a_0 > 0$, this is not possible. Thus $c_P \ne 0$ and $\|\hat{c}_P\| = 1$.

Let b' be the point defined in line 6 of Φ. If $c'^T b' = 0$, then $c^T b = 0$. Otherwise,

$$f'(b') - f'(a_0) \quad = \quad \ln \frac{(c'^T b')^n}{b'_1 \cdots b'_n} - \ln \left(\frac{c'^T a_0}{1/n} \right)^n$$

$$= \ln \left(\frac{c'^T b'}{c'^T a_0} \right)^n - \ln \left(\frac{b'_1 \cdots b'_n}{(1/n)^n} \right).$$

Lemma 54 tells us that

$$\ln \left(\frac{c'^T b'}{c'^T a_0} \right)^n \leq -\frac{\alpha}{R}.$$

This is less than $-\alpha$, since $R < 1$. Let

$$\rho = \frac{b'_1 b'_2 \cdots b'_n}{(1/n)^n} = \prod_{j=1}^{n} (nb'_j).$$

The rest of the proof consists of showing that

$$-\ln \rho = \ln \frac{1}{\rho} \leq \frac{\alpha^2}{1 - \alpha}.$$

This will imply that

$$f'(b') - f'(a_0) \leq -\alpha + \frac{\alpha^2}{1 - \alpha} = -\delta$$

and complete the proof.

Let $u = na_0 - nb' = e - nb'$; $u_j = 1 - nb'_j$ for all j. Clearly

$$\rho = \prod_{j=1}^{n} (1 - u_j). \tag{5.1}$$

Because $b' \in \Sigma$, $\sum_{j=1}^{n} u_j = 0$. Because $b' = a_0 - (\alpha/n)\hat{c}_P$, $\|a_0 - b'\| = \alpha/n$. From $\|u\| = n\|a_0 - b'\| = n \cdot (\alpha/n) = \alpha$, we infer that $\|u\|^2 = \sum_{j=1}^{n} u_j^2 = \alpha^2$. (And that $|u_j| \leq \alpha$ for all j.)

From (5.1) and the well-known fact that the geometric mean of n nonnegatives never exceeds their arithmetic mean,

$$\left(\frac{1}{\rho} \right)^{\frac{1}{n}} = \left(\prod_{j=1}^{n} \frac{1}{1 - u_j} \right)^{\frac{1}{n}} \leq \frac{1}{n} \sum_{j=1}^{n} \frac{1}{1 - u_j}.$$

So

$$\frac{1}{\rho} \le \left(\frac{1}{n} \sum_{j=1}^{n} \frac{1}{1 - u_j} \right)^n.$$

Since $|u_j| \le \alpha$, for $l \ge 2$

$$\sum_{j=1}^{n} |u_j^l| \le \sum_{j=1}^{n} \alpha^{l-2} u_j^2 = \alpha^l. \tag{5.2}$$

(This is a better bound than the obvious bound $\sum_{j=1}^{n} |u_j^l| \le n \cdot \alpha^l$.)

$$
\begin{aligned}
\sum_{j=1}^{n} \frac{1}{1 - u_j} &= \sum_{j=1}^{n} \sum_{l=0}^{\infty} u_j^l \\
&= \sum_{j=1}^{n} [\; \underset{\underset{l=0}{\uparrow}}{1} \; + \; \underset{\underset{l=1}{\uparrow}}{u_j} \; + \sum_{l=2}^{\infty} u_j^l] \\
&= n + 0 + \sum_{j=1}^{n} \sum_{l=2}^{\infty} u_j^l,
\end{aligned}
$$

because $\sum_{j=1}^{n} u_j = 0$. But

$$\sum_{j=1}^{n} \sum_{l=2}^{\infty} u_j^l \le \sum_{j=1}^{n} \sum_{l=2}^{\infty} |u_j^l| = \sum_{l=2}^{\infty} \left(\sum_{j=1}^{n} |u_j^l| \right).$$

Because $\sum_{j=1}^{n} |u_j^l| \le \alpha^l$ by (5.2),

$$\sum_{j=1}^{n} \sum_{l=2}^{\infty} u_j^l \le \sum_{l=2}^{\infty} \alpha^l = \frac{\alpha^2}{1 - \alpha}.$$

Therefore

$$\sum_{j=1}^{n} \frac{1}{1 - u_j} \le n + \frac{\alpha^2}{1 - \alpha}$$

and

$$\frac{1}{\rho} \le \left(\frac{1}{n} \left(n + \frac{\alpha^2}{1 - \alpha} \right) \right)^n = \left(1 + \frac{\alpha^2}{n(1 - \alpha)} \right)^n.$$

But $1 + \dfrac{\alpha^2}{n(1-\alpha)} \leq e^{\frac{\alpha^2}{n(1-\alpha)}}$, so that $\dfrac{1}{\rho} \leq e^{\alpha^2/(1-\alpha)}$. Therefore $\ln \dfrac{1}{\rho} \leq \dfrac{\alpha^2}{1-\alpha}$.

$$
\begin{aligned}
f'(b') - f'(a_0) &= \ln \left(\frac{c'^T b'}{c'^T a_0} \right)^n - \ln \rho \\
&\leq -\alpha + \frac{\alpha^2}{1-\alpha} = -\delta.
\end{aligned}
$$

Because $f(b) - f(a) = f'(b') - f'(a_0)$, we infer that $f(b) \leq f(a) - \delta$. ∎

Time Complexity

Each iteration requires $O(n^3)$ arithmetic operations (for the Gaussian Elimination) and the number of iterations is $O(nL)$, for a time bound of $O(n^4 L)$ arithmetic operations. Since $n^4 L$ is bounded by a polynomial in L, Karmarkar's Algorithm solves Karmarkar Standard Form problems in polynomial time. (Formally one must show that all calculations can be approximated to a polynomial number of bits of precision.)

5.4   Conversion to Karmarkar Standard Form

Now we must show how an algorithm which solves Karmarkar Standard Form linear programs in polynomial time can be used to solve arbitrary linear programs in polynomial time. It suffices to convert arbitrary linear programs to Karmarkar Standard Form.

Our jumping-off point is the Linear Inequalities problem (LI):

Find an $x \gtrless 0$ such that $Ax \leq b$, if one exists. If not, say so.

In our analysis of the Ellipsoid Algorithm, we proved that to solve arbitrary linear programs, it suffices to solve Linear Inequalities. By replacing x_j by $x_j^+ - x_j^-$ (where $x_j^+, x_j^- \geq 0$) and adding slack variables, we can convert the problem to one of the form

Find an x such that $Ax = b$, $x \geq 0$, if one exists. If not, say so.

If a linear program is feasible, it has a basic feasible solution, every entry of which is at most 2^L in absolute value (where L is the size of the problem). Suppose A is, say, $m \times r$ and let $M = r2^L$. Then we can add the constraint

$$x_1 + x_2 + \cdots + x_r \leq M$$

without changing the problem. By adding yet another slack variable x_{r+1}, we can replace that constraint by

$$x_1 + x_2 + \cdots + x_r + x_{r+1} = M.$$

Our feasibility problem is

$$\begin{aligned} Ax &= b \\ x_1 + x_2 + \cdots + x_{r+1} &= M \\ x &\geq 0, \end{aligned}$$

where A has been augmented by a column of zeroes.

If we let $x' = x/M$, then x' satisfies

$$\begin{aligned} (MA)x' &= b \\ x'_1 + x'_2 + \cdots + x'_{r+1} &= 1 \\ x' &\geq 0. \end{aligned}$$

So it suffices to find a feasible point of

$$\begin{aligned} (MA)x &= b \\ x_1 + x_2 + \cdots + x_{r+1} &= 1 \\ x &\geq 0 \end{aligned}$$

(where we've omitted the primes) and multiply the result by M. So far, so good: we have a simplex.

Subtract b_i times the constraint $e^T x = 1$ from the ith row of $(MA)x = b$, for $i = 1, 2, ..., m$. This zeroes out b, leaving us with

$$\begin{aligned} Cx &= 0 \\ x_1 + \cdots + x_{r+1} &= 1 \\ x &\geq 0 \end{aligned}$$

for some integral $m \times (r + 1)$ matrix C. Now we add a new variable λ so that a_0 will be an initial feasible point. The problem

$$\min \ \lambda$$
$$Cx - \lambda(Ce) = 0$$
$$x_1 + \cdots + x_{r+1} + \lambda = 1$$
$$x_1 \geq 0 \quad x_2 \geq 0 \quad \cdots \quad x_{r+1} \geq 0 \quad \lambda \geq 0$$

has optimal cost 0 if and only if the previous problem is feasible. The cost is always nonnegative. It can easily be verified that

$$x_1 = x_2 = \cdots = x_{r+1} = \lambda = \frac{1}{r+2}$$

is a feasible starting point. *This* is a Karmarkar Standard Form linear program on the $n = r + 2$ variables $x_1, ..., x_{r+1}, \lambda$, and the conversion is complete.

We leave it to the reader to verify that converting a problem into Karmarkar Standard Form requires only a polynomial number of operations, and that the size of the resulting problem is bounded by a polynomial in the size of the original problem. (This is not the most "efficient" conversion scheme. Other methods lead to smaller Karmarkar Standard Form linear programs.)

Theorem 56. Karmarkar's Algorithm solves arbitrary linear programs in polynomial time on a RAM that performs exact arithmetic.

In fact, Karmarkar's Algorithm runs in polynomial time even on a traditional limited-precision RAM.

■ Karmarkar's Algorithm

5.5 Notes

The jury is still out on whether Karmarkar's Algorithm is a practical improvement over Simplex. Some researchers have reported results much less impressive than Karmarkar's claim that his algorithm often outperforms Simplex by a factor of 50 on real-world problems. Experimental results on the performance of Karmarkar's Algorithm appear in Adler, Resende and Veiga [5], Goldfarb and Mehrotra [24], and Tomlin [59]. Our coverage of Karmarkar's Algorithm is based primarily on Karmarkar's original paper [34], and slightly on the notes by Babai [10].

The polynomial-time linear programming algorithms of Renegar [45], Monteiro and Adler [42], Vaidya [60] and Gonzaga [26] postdate Karmarkar's Algorithm and have better worst-case running time bounds, at least for certain ranges of m and n.

The reader can find a comparison of the Ellipsoid Algorithm and Karmarkar's Algorithm in Todd [58]. Goldfarb and Todd's superb 98-page article, "Linear Programming" [25], discusses the Simplex Algorithm, duality, and the Ellipsoid Algorithm, as well as Karmarkar's Algorithm and several later polynomial-time algorithms.

Karmarkar's Algorithm can be proven to require only polynomial precision via a proof similar to that used for the Ellipsoid Algorithm (see the comments in Schrijver [47] and Renegar [45]). Also, the ideas in Vaidya's analysis of the precision required by his linear programming algorithm [60] can be used to bound the precision required by Karmarkar's Algorithm. Surprisingly, no proof expressly for Karmarkar's Algorithm has been published.

Bibliography

[1] I. Adler, "The Expected Number of Pivots Needed to Solve Parametric Linear Programs and the Efficiency of the Self-Dual Simplex Method," manuscript, Dept. of Industrial Engineering and Operations Research, University of California, Berkeley, 1983.

[2] I. Adler, R. M. Karp and R. Shamir, "A Family of Simplex Variants Solving an $m \times d$ Linear Program in Expected Number of Pivots Depending on d Only," *Mathematics of Operations Research* 11 (1986), 570-590.

[3] I. Adler, R. M. Karp and R. Shamir, "A Simplex Variant Solving an $m \times d$ Linear Program in $O(\min(m^2, d^2))$ Expected Number of Pivot Steps," *Journal of Complexity* 3 (1987), 372-387.

[4] I. Adler and N. Megiddo, "A Simplex Algorithm Whose Average Number of Steps is Bounded Between Two Quadratic Functions of the Smaller Dimension," *Journal of the ACM* 32 (1985), 871-895.

[5] I. Adler, M. G. C. Resende and G. Veiga, "An Implementation of Karmarkar's Algorithm for Linear Programming," Report ORC 86-8, Department of Industrial Engineering and Operations Research, University of California, Berkeley, 1986.

[6] A. V. Aho, J. E. Hopcroft and J. D. Ullman, *The Design and Analysis of Computer Algorithms*, Addison-Wesley, Reading, MA, 1974.

[7] B. Aspvall and R. E. Stone, "Khachiyan's Linear Programming Algorithm," *Journal of Algorithms* 1 (1980), 1-13.

[8] D. Avis and V. Chvátal, "Notes on Bland's Pivoting Rule," *Mathematical Programming Study* 8 (1978), 24-34.

[9] S. Baase, *Computer Algorithms: Introduction to Design and Analysis*, second edition, Addison-Wesley, Reading, MA, 1988.

[10] L. Babai, class notes, Dept. of Computer Science, University of Chicago, Chicago, IL.

[11] R. G. Bland, "New Finite Pivoting Rules for the Simplex Method," *Mathematics of Operations Research* 2 (1977), 103-107.

[12] R. G. Bland, D. Goldfarb and M. J. Todd, "The Ellipsoid Method: A Survey," *Operations Research* 29 (1981), 1039-1091.

[13] K. H. Borgwardt, *The Simplex Method: A Probabilistic Analysis*, Springer-Verlag, New York, NY, 1987.

[14] G. Brassard and P. Bratley, *Algorithmics: Theory and Practice*, Prentice Hall, Englewood Cliffs, NJ, 1988.

[15] J. E. Calvert and W. L. Voxman, *Linear Programming*, Harcourt Brace Jovanovich, New York, NY, 1989.

[16] V. Chvátal, *Linear Programming*, Freeman, New York, NY, 1983.

[17] G. B. Dantzig, *Linear Programming and Extensions*, Princeton University Press, Princeton, NJ, 1963.

[18] J. Edmonds, "Systems of Distinct Representatives and Linear Algebra," *Journal of Research of the National Bureau of Standards B* 71 (1967), 241-245.

[19] P. Gács and L. Lovász, "Khachiyan's Algorithm for Linear Programming," *Mathematical Programming Study 14* (1981), 61-68.

[20] D. Gale, *The Theory of Linear Economic Models*, University of Chicago Press, Chicago, IL, 1988.

[21] D. Gale, H. W. Kuhn and A. W. Tucker, "On Symmetric Games," in *Contributions to the Theory of Games*, H. W. Kuhn and A. W. Tucker, eds., *Ann. Math Studies* no. 24, Princeton University Press, Princeton, NJ, 1950.

[22] S. I. Gass, *Linear Programming Methods and Applications*, fourth edition, McGraw-Hill, New York, NY, 1975.

[23] A. M. Glicksman, *An Introduction to Linear Programming and the Theory of Games*, Wiley, New York, NY, 1963.

[24] D. Goldfarb and S. Mehrotra, "A Relaxed Version of Karmarkar's Algorithm," *Mathematical Programming* 40 (1988), 289-315.

[25] D. Goldfarb and M. J. Todd, "Linear Programming," in *Optimization*, G. Nemhauser et al., eds., North-Holland, Amsterdam, 1989, 73-170.

[26] C. C. Gonzaga, "An Algorithm for Solving Linear Programming Problems in $O(n^3 L)$ Operations," in *Progress in Mathematical Programming: Interior-Point and Related Methods*, N. Megiddo, ed., Springer-Verlag, New York, NY, 1989, 1-28.

[27] M. Grötschel, L. Lovász and A. Schrijver, *Geometric Algorithms and Combinatorial Optimization*, Springer-Verlag, New York, NY, 1988.

[28] G. Hadley, *Linear Programming*, Addison-Wesley, Reading, MA, 1962.

[29] I. Herstein, *Topics in Algebra*, second edition, Wiley, New York, NY, 1975.

[30] A. J. Hoffman, "Cycling in the Simplex Algorithm," Report No. 2974, National Bureau of Standards, Washington, DC, 1953.

[31] K. Hoffman and R. Kunze, *Linear Algebra*, Prentice Hall, Englewood Cliffs, NJ, 1971.

[32] R. G. Jeroslow, "The Simplex Algorithm with the Pivot Rule of Maximizing Criterion Improvement," *Discrete Mathematics* 4 (1973), 367-377.

[33] D. B. Judin and A. S. Nemirovskii, "Informational Complexity and Effective Methods for the Solution of Convex Extremal Problems" (in Russian), *Ekonomika; Matematicheskie Metody* 12 (1976), 357-369.

[34] N. Karmarkar, "A New Polynomial-Time Algorithm for Linear Programming," *Combinatorica* 4 (1984), 373-395.

[35] L. G. Khachiyan, "A Polynomial Algorithm for Linear Programming" (in Russian), *Doklady Akademiia Nauk USSR* 244 (1979), 1093-1096. A translation appears in *Soviet Mathematics Doklady* 20 (1979), 191-194.

[36] V. Klee and G. J. Minty, "How Good is the Simplex Algorithm?", in *Inequalities-III*, O. Shisha, ed., Academic Press, New York, NY, 1972, 159-175.

[37] H. W. Kuhn, class notes, Princeton University, Princeton, NJ, 1976.

[38] H. W. Kuhn, "Solvability and Consistency for Linear Equations and Inequalities," *American Mathematics Monthly* 63 (1956), 217-232.

[39] E. L. Lawler, *Combinatorial Optimization: Networks and Matroids*, Holt, Rinehart and Winston, New York, NY, 1976.

[40] E. L. Lawler, "The Great Mathematical Sputnik of 1979," *The Mathematical Intelligencer* 2 (1980), 191-198.

[41] U. Manber, *Introduction to Algorithms: A Creative Approach*, Addison-Wesley, Reading, MA, 1989.

[42] R. D. C. Monteiro and I. Adler, "Interior Path Following Primal-Dual Algorithms. Part I: Linear Programming," *Mathematical Programming* 44 (1989), 27-41.

[43] G. Owen, *Game Theory*, second edition, Academic Press, New York, NY, 1982.

[44] C. H. Papadimitriou and K. Steiglitz, *Combinatorial Optimization: Algorithms and Complexity*, Prentice Hall, Englewood Cliffs, NJ, 1982.

[45] J. Renegar, "A Polynomial-Time Algorithm, Based on Newton's Method, for Linear Programming," *Mathematical Programming* 40 (1988), 50-94.

[46] W. Rudin, *Real and Complex Analysis*, third edition, McGraw-Hill, New York, NY, 1987, 54-55.

[47] A. Schrijver, *Theory of Linear and Integer Programming*, Wiley, New York, NY, 1986.

[48] R. Shamir, "The Efficiency of the Simplex Method: A Survey," *Management Science* 33 (1987), 301-334.

[49] L. S. Shapley, *Theory of Games and Its Applications to Economics and Politics*, Macmillan India, Delhi, 1981.

[50] N. Z. Shor, "Utilization of the Operation of Space Dilation in the Minimization of Convex Functions" (in Russian), *Kibernetika* 1 (1970), 6-12. A translation appears in *Cybernetics* 6 (1970), 7-15.

[51] M. Shubik, *Game Theory in the Social Sciences: Concepts and Solutions*, MIT Press, Cambridge, MA, 1982.

[52] M. Shubik, *A Game-Theoretic Approach to Political Economy*, MIT Press, Cambridge, MA, 1984.

[53] M. Simonnard, *Linear Programming*, Prentice Hall, Englewood Cliffs, NJ, 1966.

[54] S. Smale, "The Problem of the Average Speed of the Simplex Method," in *Mathematical Programming: The State of the Art (Bonn 1982)*, A. Bachem et al., eds., Springer-Verlag, Berlin, 1983, 530-539.

[55] S. Smale, "On the Average Number of Steps in the Simplex Method of Linear Programming," *Mathematical Programming* 27 (1983), 241-262.

[56] G. Strang, *Linear Algebra and Its Applications*, third edition, Harcourt Brace Jovanovich, San Diego, CA, 1988.

[57] É. Tardos, "A Strongly Polynomial Algorithm to Solve Combinatorial Linear Programs," *Operations Research* 34 (1986), 250-256.

[58] M. J. Todd, "Polynomial Algorithms for Linear Programming," in *Advances in Optimization and Control*, H. A. Eiselt and G. Pederzoli, eds., Springer-Verlag, Berlin, 1988, 49-66.

[59] J. A. Tomlin, "An Experimental Approach to Karmarkar's Projective Method for Linear Programming," *Mathematical Programming Study* 31 (1987), 175-191.

[60] P. M. Vaidya, "An Algorithm for Linear Programming Which Requires $O(((m+n)n^2 + (m+n)^{1.5}n)L)$ Arithmetic Operations," *Mathematical Programming* 47 (1990), 175-201.

[61] J. von Neumann and O. Morgenstern, *Theory of Games and Economic Behavior*, third edition, Princeton University Press, Princeton, New Jersey, 1980.

Index

Progress in Theoretical Computer Science

Progress in Theoretical Computer Science is a series that focuses on the theoretical aspects of computer science and on the logical and mathematical foundations of computer science, as well as the applications of computer theory. It addresses itself to research workers and graduate students in computer and information science departments and research laboratories, as well as to departments of mathematics and electrical engineering where an interest in computer theory is found.

The series publishes research monographs, graduate texts, and polished lectures from seminars and lecture series. We encourage preparation of manuscripts in some form of TeX for delivery in camera-ready copy, which leads to rapid publication, or in electronic form for interfacing with laser printers or typesetters.

Proposals should be sent directly to the Editor, any member of the Editorial Board, or to: Birkhäuser Boston, 675 Massachusetts Avenue, Cambridge, MA 02139.